Harald Held

Shape Optimization under Uncertainty
from a Stochastic Programming Point of View

VIEWEG+TEUBNER RESEARCH

Stochastic Programming

Editor:
Prof. Dr. Rüdiger Schultz

Uncertainty is a prevailing issue in a growing number of optimization problems in science, engineering, and economics. Stochastic programming offers a flexible methodology for mathematical optimization problems involving uncertain parameters for which probabilistic information is available. This covers model formulation, model analysis, numerical solution methods, and practical implementations. The series "Stochastic Programming" presents original research from this range of topics.

Harald Held

Shape Optimization under Uncertainty from a Stochastic Programming Point of View

With a foreword by Prof. Dr. Rüdiger Schultz

VIEWEG+TEUBNER RESEARCH

Bibliografische Information der Deutschen Nationalbibliothek
Die Deutsche Nationalbibliothek verzeichnet diese Publikation in der
Deutschen Nationalbibliografie; detaillierte bibliografische Daten sind im Internet über
<http://dnb.d-nb.de> abrufbar.

Beim vorliegenden Buch handelt es sich um eine vom Fachbereich Mathematik
der Universität Duisburg-Essen genehmigte Dissertation.

Datum der mündlichen Prüfung: 02. Februar 2009

Referent: Prof. Dr. Rüdiger Schultz
Korreferent: Prof. Dr. Martin Rumpf

1. Auflage 2009

Alle Rechte vorbehalten
© Vieweg+Teubner | GWV Fachverlage GmbH, Wiesbaden 2009

Lektorat: Christel A. Roß | Anita Wilke

Vieweg+Teubner ist Teil der Fachverlagsgruppe Springer Science+Business Media.
www.viewegteubner.de

Das Werk einschließlich aller seiner Teile ist urheberrechtlich geschützt.
Jede Verwertung außerhalb der engen Grenzen des Urheberrechtsgesetzes ist ohne Zustimmung des Verlags unzulässig und strafbar. Das gilt
insbesondere für Vervielfältigungen, Übersetzungen, Mikroverfilmungen
und die Einspeicherung und Verarbeitung in elektronischen Systemen.

Die Wiedergabe von Gebrauchsnamen, Handelsnamen, Warenbezeichnungen usw. in diesem
Werk berechtigt auch ohne besondere Kennzeichnung nicht zu der Annahme, dass solche
Namen im Sinne der Warenzeichen- und Markenschutz-Gesetzgebung als frei zu betrachten
wären und daher von jedermann benutzt werden dürften.

Umschlaggestaltung: KünkelLopka Medienentwicklung, Heidelberg
Druck und buchbinderische Verarbeitung: STRAUSS GMBH, Mörlenbach
Gedruckt auf säurefreiem und chlorfrei gebleichtem Papier.
Printed in Germany

ISBN 978-3-8348-0909-4

Foreword

Optimization problems whose constraints involve partial differential equations (PDEs) are relevant in many areas of technical, industrial, and economic applications. At the same time, they pose challenging mathematical research problems in numerical analysis and optimization.

The present text is among the first in the research literature addressing stochastic uncertainty in the context of PDE constrained optimization. The focus is on shape optimization for elastic bodies under stochastic loading. Analogies to finite dimensional two-stage stochastic programming drive the treatment, with shapes taking the role of nonanticipative decisions. The main results concern level set-based stochastic shape optimization with gradient methods involving shape and topological derivatives. The special structure of the elasticity PDE enables the numerical solution of stochastic shape optimization problems with an arbitrary number of scenarios without increasing the computational effort significantly. Both risk neutral and risk averse models are investigated.

This monograph is based on a doctoral dissertation prepared during 2004-2008 at the Chair of Discrete Mathematics and Optimization in the Department of Mathematics of the University of Duisburg-Essen. The work was supported by the Deutsche Forschungsgemeinschaft (DFG) within the Priority Program "Optimization with Partial Differential Equations".

Rüdiger Schultz

Acknowledgments

I owe a great deal to my supervisors, colleagues, and friends who have always supported, encouraged, and enlightened me through their own research, comments, and questions.

When I started as a freshman at the University of Duisburg nearly a decade ago, the very first lecture I attended was given by Prof. Dr. Rüdiger Schultz. Undoubtedly, it was his enthusiasm and passion for mathematics that kindled my interests and ambitions at that time, which finally led to this thesis. For that, his constant motivation and invaluable advice, his encouragement to pursue my own ideas, and the faith he put in me, I thank him deeply.

I further thank Prof. Dr. Martin Rumpf for his support, invaluable advice, and helpful ideas that proved useful in many difficult situations. I am also thankful to him and his group for allowing me access to their excellent software library, which was a great asset to my research.

I am grateful to all of my colleagues for their willingness to genuinely help and discuss virtually everything at any time, providing the most pleasant work environment. In particular, I would like to express my gratitude to Ralf Gollmer, Uwe Gotzes, and Martin Pach for fruitful discussions and suggestions.

Not least I thank my wife, Karina, for her patience, love, and proofreading. I spent many evenings and weekends writing and "bug squishing". For that I am in her debt.

Harald Held

Abstract

We consider an elastic body subjected to internal and external forces which are uncertain. Simply averaging the possible loadings will result in a structure that might not be robust for the individual loadings at all. Instead, we apply techniques from level set-based shape optimization and two-stage stochastic programming: In the first stage, the non-anticipative decision on the shape has to be taken. Afterwards, the realizations of the random forces are observed, and the variational formulation of the elasticity system takes the role of the second-stage problem. Taking advantage of the PDE's linearity, we are able to compute solutions for an arbitrary number of scenarios without increasing the computational effort significantly. The deformations are described by PDEs that are solved efficiently by Composite Finite Elements. The objective is, e.g., to minimize the compliance. A gradient method using the shape derivative is used to solve the problem. Results for 2D instances are shown. The obtained solutions strongly depend on the initial guess, in particular its topology. To overcome this issue, we included the topological derivative into our algorithm as well.

The stochastic programming perspective also allows us to incorporate risk measures into our model which might be a more appropriate objective in many practical applications.

Parts of this work have been published in [CHP$^+$09].

Contents

1 **Introduction**   1
   1.1 The Elasticity PDE 4
   1.2 Shape Optimization Problems 13
   1.3 Two-Stage Stochastic Programming 17

2 **Solution of the Elasticity PDE**   25
   2.1 Composite Finite Elements 28

3 **Stochastic Programming Perspective**   49
   3.1 Stochastic Shape Optimization Problem 50
   3.2 Reformulation and Solution Plan for the Expectation-Based Model   61
   3.3 Expected Excess 70
   3.4 Excess Probability 73

4 **Solving Shape Optimization Problems**   77
   4.1 Level Set Formulation 78
   4.2 Shape Derivative 81
   4.3 Topological Derivative 89
   4.4 Steepest Descent Algorithm 94

5 **Numerical Results**   101
   5.1 Deterministic and Expectation-Based Results 102
   5.2 Risk Aversion 114

A **Appendix**   121
   A.1 Notation 121
   A.2 Important Facts and Theorems 124

Appendix   121

References   127

Symbol Index

$\mathcal{O}$	The elastic body
Γ_0	Part of the boundary that is to be optimized
Γ_D	The fixed Dirichlet boundary
Γ_N	Neumann boundary where the surface loads act on
λ, μ	Lamé coefficients
ϕ	Level set function
π	Vector of probabilities
$\mathbb{R}^n$	n-dimensional Euclidean space
ω	A scenario
$\mathbf{J}(\mathcal{O}) = J(\mathcal{O}, u(\mathcal{O}))$	Shape objective functional
A	Elasticity tensor
$e(u)$	Linearized strain tensor
$f_{,i}$	i^{th} partial derivative of a scalar function f, see A.4 on page 122
V	Function space $H^1_{\Gamma_D}(\mathcal{O}; \mathbb{R}^2)$
D	Working domain that contains all admissible shapes

A more detailed overview of the notations we used can be found in the Appendix A on page 121.

1 Introduction

Shape optimization problems arise in various practical applications. As stated in [DJPZ01], the object that is to be optimized is the *geometry* as a variable. Shape optimization is closely related to topology optimization, where not only the shape and sizing of a structure has to be found, but also the topology, i.e. the location and shape of holes (see e.g. [BS03]).

In this work, we consider an elastic body represented by an open bounded domain[1] $\mathscr{O} \subset \mathbb{R}^2$. This elastic body is subjected to volume forces and surface loads which are unknown in advance and vary stochastically over time. The objective is to find a shape that minimizes a certain functional under the given loading conditions. Of course, since the acting forces are uncertain and therefore not known in advance, one has to decide on the shape before one can observe the actual forces. This resembles the ideas and structure of linear two-stage stochastic programming problems. This work works out this analogy in the case of shape optimization for linear elastic material laws and stochastic volume and surface loadings.

The motivation behind the stochastic approach becomes evident when looking at the following particular situation, which is also described in [CC03]: Suppose, our task is to find a design for some elastic mechanical device that is as stiff as possible. The stiffness that is to be maximized in this context is an elastic energy as the result of applying forces acting on the design. Under the assumption that the loading is fixed and known, the optimization process yields a structure which resists that one particular given force as good as possible. It is not difficult to imagine situations where the optimal design is unstable with respect to variations of the forces. See for example instance Fig. 5.2 on page 103 in Chapter 5. There we have a square supported on its bottom edge and a homogeneous vertical surface load is acting on its upper edge. The resulting optimal structure consists of vertical pillars (see Fig. 5.2 (left)), which is clearly not optimal any more for any other but the given vertical loading. Note that the instability is not a malfunction in the optimization procedure but the model itself. One can only hope to find more stable and robust solutions if the model somehow incorporates uncertain loadings.

---

[1] Note that all results described here also hold for $\mathscr{O} \subset \mathbb{R}^3$. However, the computational results are all obtained for the 2-dimensional case, so for the ease of presentation we restrict ourselves to $\mathbb{R}^2$.

One way to achieve this is the stochastic programming approach to this kind of problem, which is the main contribution of this work.

Another possibility to avoid the vulnerability of the optimal designs with respect to variations of loadings, is the *robust optimization* approach. For details about robust optimization we refer to Ben-Tal et al. [BTN02] and references therein, here we only state the basic idea. Robust optimization aims to solve optimization problems in which some data are uncertain and is only known to belong to some uncertainty set $\mathscr{U}$. The following general (finite dimensional) optimization problem is considered in [BTN02]:

$$\min_{x_0 \in \mathbb{R}, x \in \mathbb{R}^n} \{x_0 : f_0(x,\zeta) - x_0 \leq 0, \quad f_i(x,\zeta) \leq 0, \quad i=1,\ldots,m\} \quad (1.1)$$

with the design vector x, the objective function f_0, constraints $f_1,\ldots,f_m$, and uncertain data $\zeta \in \mathscr{U}$. Then, one associates with the uncertain problem (1.1) its so-called *robust counterpart* which is the (semi-infinite) optimization problem

$$\min_{x_0,x} \{x_0 : f_0(x,\zeta) \leq x_0, \quad f_i(x,\zeta) \leq 0, \quad i=1,\ldots,m \quad \forall \zeta \in \mathscr{U}\}. \quad (1.2)$$

Note that in particular any feasible x and x_0 in (1.2) have to satisfy the constraint $f_0(x,\zeta) \leq x_0, \forall \zeta \in \mathscr{U}$, which can be stated equivalently as $\max_{\zeta \in \mathscr{U}} f_0(x,\zeta) \leq x_0$. The right-hand side x_0 is the objective function in (1.2) which is to be minimized. Consequently, for an optimal design vector $\bar{x}$ we have $x_0 = \max_{\zeta \in \mathscr{U}} f_0(\bar{x},\zeta)$. In this sense, the robust counterpart (1.2) overcomes the issue of instability due to uncertain data by minimizing the worst possible case in the given range of data.

The idea of robust optimization has been applied to practical shape and topology optimization applications, such as airfoil shape optimization for example, where the forces are not always known in advance and may vary intensely. This is carried out for example in [Huy01]. Other applications and model formulations for robust shape optimization problems can be found e.g. in [CC99, CC03, dGAJ06]. To our knowledge, the ideas of stochastic (two-stage) programming, which also take the distribution of the random data into account, have not been applied to shape optimization problems under uncertainty yet.

In Section 1.2 we give an introduction to deterministic shape optimization problems. Section 1.1 deals with the formulation and properties of the underlying elasticity PDE[2]. The introduction closes with the ideas and important concepts of two-stage stochastic programming in Section 1.3.

Chapter 2 describes in detail the finite element method we used — the so-called Composite Finite Elements — to solve the elasticity PDE, including some implementational details.

---

[2]Partial Differential Equation

1 Introduction

In Chapter 3 we show how some ideas from finite dimensional two-stage stochastic programming can be applied to the infinite dimensional setting of our stochastic shape optimization problems. It turns out that for this purpose duality plays an important role for an efficient way to compute solutions. This is worked out in Section 3.1. A reformulation of the stochastic shape optimization problem which suggests an immediate way to evaluate the objective function is obtained in Section 3.2. Based on this formulation of the problem, risk averse objective functionals are quite easy to be included, which can be found in Sections 3.3 and 3.4.

Of course, after having formulated appropriate stochastic shape optimization problems, one is also interested in solving them numerically. Along with this work, we developed a program which does that for the 2-dimensional case. The algorithm we implemented is essentially a steepest descent algorithm combined with a level set method. We mainly follow [AJT04] in that respect. In Section 4.1 we describe how we represent domains via level set functions, and what properties and advantages level set methods have. As mentioned before, we want to apply a steepest descent algorithm, so we need to know how to evaluate the objective function, and how to compute a descent direction. The former becomes clear in Chapter 3, and the latter is dealt with in Chapter 4. In particular, in Section 4.2 the notion of *shape derivative* is introduced which is essential for computing a descent direction.

One drawback of a steepest descent algorithm for our problem is that it requires an initial guess. In other words, one has to decide on a certain topology[3]. It turns out that this has a great influence on the outcome of the optimization algorithm (see e.g. [AJT04, AdGJT05, BS03]). The notion of convexity does not apply for functionals depending on domains. Hence there is no guarantee that a steepest descent algorithm finds an optimal solution. In general, one can only say that it terminates in a critical point (cf. for example [BGLS03, NW99, Rus06]). Moreover, the used level set method is in general not able to create new holes (see [AJT04]) but might be able to join several holes together. One attempt to overcome those problems is to embed the *topological derivative* as e.g. in [AdGJT05, BHR04]. More on the topological derivative and topology optimization in general can be found for instance in [AdGJT05, BS03, BHR04, BO05, GGM01, HL07, SZ99, SZ01] and references therein. We also included the topological derivative in our implementation which is described in Section 4.3. Finally, the complete algorithm is summarized and presented in Section 4.4.

Numerical results for the 2-dimensional case are presented in Chapter 5. For convenience, we summarized all the notations we used in the Appendix A.1.

---

[3] Here we mean the number of holes and their size and location.

1.1 The Elasticity PDE

As mentioned before, we seek to optimize the shape of an elastic body $\mathcal{O} \subseteq \mathbb{R}^2$ subjected to internal and external forces. Here we only want to give a brief introduction to elasticity and the PDE which serves as the state equation for the shape optimization problems that are considered in this work. More on elasticity theory can be found in [Cia88] and [Bra03]. The latter also addresses computational aspects using finite element methods.

Due to the forces acting on the body $\mathcal{O}$, the body is deformed and a point $x \in \mathcal{O}$ becomes the point x' of the deformed body as illustrated in Figure 1.1. Then we can express x' as $x' = x + u(x)$, where $u \colon \mathbb{R}^2 \to \mathbb{R}^2$ denotes the vector of displacement and is assumed to be sufficiently smooth. Those displacements are often assumed to be small and thus higher order terms in u are neglected. This leads to the theory of linearized elasticity which we consider in this work for isotropic elastic materials. One of the most important notions in elasticity theory is the *strain tensor*

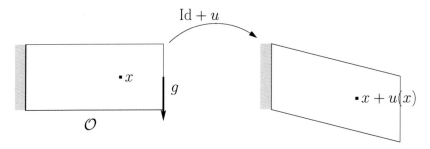

Fig. 1.1: Sketch of an elastic body $\mathcal{O}$ which is fixed on its left edge. Due to the surface load g the body deforms, and a point $x \in \mathcal{O}$ becomes $x + u(x)$.

which reads in the linearized theory as[4]

$$e_{ij}(u) := \frac{1}{2}\left(u_{i,j} + u_{j,i}\right). \tag{1.3}$$

The 2×2 matrix $e(u) = (e_{ij}(u))$ is obviously symmetric, and the mapping $u \mapsto e(u)$ linear.

We distinguish between *volume forces f* and *surface loads g*. A typical example for a volume force is gravity, whereas an imposed load on a bridge would be a surface load. The resulting deformation due to those forces obviously depends on

---
[4]For the notation we used here for derivatives, see A.1, in particular A.4 (ii)

1.1 The Elasticity PDE

the material the elastic body is made of. Here we consider a linearized material law for isotropic elastic material, known as Hooke's law (see Definition 1.1).

Definition 1.1. *Let $i, j, k, l \in \{1, 2\}$ be indices. For notations see A.2.*

(i) *The elasticity tensor $A = (A_{ijkl})_{ijkl}$ is defined by*

$$A_{ijkl} = 2\mu \delta_{ik}\delta_{jl} + \lambda \delta_{ij}\delta_{kl}. \tag{1.4}$$

Note that $A_{ijkl} = A_{klij} = A_{jilk}$.

(ii) *$\sigma = (\sigma_{ij})_{ij}$ with $\sigma_{ij} := \sum_{k,l} A_{ijkl} e_{kl}(u) = \sigma_{ji}$ is called* **Cauchy stress tensor** *and constitutes Hooke's law.*

(iii) *λ and μ are material constants called* **Lamé** *coefficients. According to [Cia88], those constants of actual materials are greater than 0.*

Note that for any symmetric 2×2 matrix $\xi = (\xi_{ij})$ we have

$$\sum_{k,l} A_{ijkl}\xi_{kl} = \sum_{k,l} 2\mu \delta_{ik}\delta_{jl}\xi_{kl} + \lambda \delta_{ij}\delta_{kl}\xi_{kl}$$
$$= \sum_{k} 2\mu \delta_{ik}\xi_{kj} + \lambda \delta_{ij}\xi_{kk}$$
$$= 2\mu \xi_{ij} + \lambda \delta_{ij}\operatorname{tr}(\xi),$$

and therefore we obtain

$$A\xi = 2\mu \xi + \lambda \operatorname{tr}(\xi)\operatorname{Id}. \tag{1.5}$$

Throughout this work, we assume that the domain $\mathscr{O}$ has Lipschitz boundary[5]. We further always assume the following configuration: $\partial \mathscr{O}$ consists of three disjoint parts

$$\partial \mathscr{O} = \Gamma_D \cup \Gamma_N \cup \Gamma_0, \tag{1.6}$$

which have the following properties:

- Γ_D is the fixed **Dirichlet boundary**, i.e. the displacement u is required to be 0 on Γ_D. We assume that Γ_D is not allowed to move during the optimization process. Moreover, we require that $\Gamma_D \neq \emptyset$[6].

- Γ_N is the part of the boundary where the surface loads g act on. We also require this part of the boundary of $\mathscr{O}$ to be fixed such that it does not move during the optimization process.

---
[5] See e.g. [Alt02, p. 242] for a definition of Lipschitz boundary
[6] Physically this assumption makes sense because without it the resulting displacement would be infinite. Also analytically it is needed to ensure the existence of a unique solution u to (1.7).

- Γ_0 consists of the remaining part of $\partial\mathcal{O}$, i.e. $\Gamma_0 = \partial\mathcal{O}\setminus\Gamma_D\setminus\Gamma_N$. Because of the assumptions that Γ_D and Γ_N are fixed during the optimization process, this is the only part of $\partial\mathcal{O}$ to be optimized.

The displacement $u\colon \mathcal{O} \to \mathbb{R}^2$ is then determined as the solution to the following system of linear partial differential equations:

$$\begin{cases} -\operatorname{div}(Ae(u)) = f & \text{in } \mathcal{O}, \\ u = 0 & \text{on } \Gamma_D, \\ (Ae(u))n = g & \text{on } \Gamma_N, \\ (Ae(u))n = 0 & \text{on } \Gamma_0, \end{cases} \tag{1.7}$$

where n denotes the *outward pointing* unit normal vector field along $\partial\mathcal{O}$. With Definition 1.1 and Notation A.4 on page 122, the first line in (1.7) is to be understood as

$$-\sum_{j=1}^{2}\sigma_{ij,j} = f_i, \quad \text{for } i = 1,2.$$

The algorithm we employ to solve our shape optimization problems is a steepest descent method (cf. Section 4.4), and as such computes different shapes (i.e. domains $\mathcal{O}$) in each iteration. Due to the varying of $\mathcal{O}$ during the optimization process, the forces f and g must be known for all possible configurations of $\mathcal{O}$. For that purpose we introduce a working domain $D \subseteq \mathbb{R}^2$ that contains all admissible shapes $\mathcal{O}$ (cf. Fig. 1.2). Without loss of generality, we can always assume that $D = (0,1) \times (0,1)$ — this can always be achieved by scaling appropriately. Consequently, we suppose that $f \in L^2(D;\mathbb{R}^2)$ and $g \in H^1(D;\mathbb{R}^2)$. Since $\mathcal{O}$ has Lipschitz boundary, g has traces on $\partial\mathcal{O}$ in $L^2(\partial\mathcal{O};\mathbb{R}^2)$ (cf. [Alt02, p. 249]).

1.1.1 Variational Formulation

In this section we show that a solution to (1.7) can be equivalently characterized as a minimizer of a quadratic variational problem. Most presented proofs here follow the ideas described in [Cia88]. The domain $\mathcal{O}$ is held fixed throughout this section.

Definition 1.2. *In the sequel we denote by V the function space*

$$V := H^1_{\Gamma_D}(\mathcal{O};\mathbb{R}^2) = \left\{u \in H^1(\mathcal{O};\mathbb{R}^2) : u = 0 \text{ on } \Gamma_D \text{ in the sense of traces}\right\}.$$

We further define for $u, \psi, \vartheta \in V$

$$E(\mathcal{O},u) := \frac{1}{2}A(\mathcal{O},u,u) - l(\mathcal{O},u) \tag{1.8}$$

1.1 The Elasticity PDE

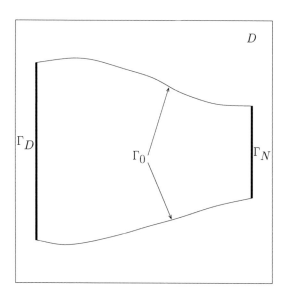

Fig. 1.2: A sketch of an admissible domain with the boundary configuration.

with

$$A(\mathcal{O}, \psi, \vartheta) := \int_{\mathcal{O}} \sum_{i,j,k,l} A_{ijkl} e_{ij}(\psi) e_{kl}(\vartheta) \, dx = \int_{\mathcal{O}} Ae(\psi) : e(\vartheta) \, dx, \qquad (1.9)$$

$$l(\mathcal{O}, \vartheta) := \int_{\mathcal{O}} f \cdot \vartheta \, dx + \int_{\Gamma_N} g \cdot \vartheta \, ds. \qquad (1.10)$$

The following Lemma summarizes some simple but important facts.

Lemma 1.3.

(i) *For all symmetric matrices $\xi \in \mathbb{R}^{2 \times 2}$ it holds that $A\xi : \xi \geq 2\mu \xi : \xi$.*

(ii) $Ae(\psi) : e(\vartheta) = Ae(\vartheta) : e(\psi), \quad \forall \vartheta, \psi \in V$

(iii) $\frac{d}{d\varepsilon} \left(Ae(\psi + \varepsilon \varphi) : e(\psi + \varepsilon \varphi) \right)\big|_{\varepsilon=0} = 2Ae(\psi) : e(\varphi), \quad \forall \psi, \varphi \in V.$

(iv) *The bilinear form $A(\mathcal{O}, \psi, \vartheta)$ is V-elliptic in the sense that there is a constant $\alpha > 0$ such that $A(\mathcal{O}, v, v) \geq \alpha \|v\|_1^2, \quad \forall v \in V.$*

(v) *$l(\mathcal{O}, \cdot)$ is a continuous linear form on V.*

(vi) The bilinear form (1.9) is bounded, i.e. there exists a constant C such that

$$|A(\mathscr{O},\psi,\vartheta)| \leq C\|\psi\|_1\|\vartheta\|_1 \text{ for all } \psi,\vartheta \in V.$$

Proof.

(i) As noted in Definition 1.1, we have $\lambda, \mu > 0$. Using (1.5) yields

$$\begin{aligned} A\xi : \xi &= (2\mu\xi + \lambda \operatorname{tr}(\xi)\operatorname{Id}) : \xi \\ &= 2\mu\xi : \xi + \lambda \operatorname{tr}(\xi)\operatorname{Id} : \xi \\ &= 2\mu\xi : \xi + \lambda \left(\operatorname{tr}(\xi)\right)^2 \\ &\geq 2\mu\xi : \xi. \end{aligned}$$

(ii) Due to the symmetry of A_{ijkl} noted in Definition 1.1, it holds that

$$\begin{aligned} Ae(\psi) : e(\vartheta) &= \sum_{i,j,k,l} A_{ijkl} e_{kl}(\psi) e_{ij}(\vartheta) \\ &= \sum_{i,j,k,l} A_{klij} e_{ij}(\vartheta) e_{kl}(\psi) \\ &= Ae(\vartheta) : e(\psi). \end{aligned}$$

(iii) Because of the linearity of the mapping $u \mapsto e(u)$ we have

$$\begin{aligned} Ae(\psi+\varepsilon\varphi) : e(\psi+\varepsilon\varphi) &= \sum_{i,j,k,l} A_{ijkl} e_{kl}(\psi+\varepsilon\varphi) e_{ij}(\psi+\varepsilon\varphi) \\ &= \sum_{i,j,k,l} \left[A_{ijkl} e_{kl}(\psi) e_{ij}(\psi) + \varepsilon A_{ijkl} e_{kl}(\varphi) e_{ij}(\psi) \right. \\ &\quad \left. + \varepsilon A_{ijkl} e_{kl}(\psi) e_{ij}(\varphi) + \varepsilon^2 A_{ijkl} e_{kl}(\varphi) e_{ij}(\varphi)\right] \\ &= Ae(\psi) : e(\psi) + \varepsilon \left(Ae(\varphi) : e(\psi) + Ae(\psi) : e(\varphi)\right) \\ &\quad + \varepsilon^2 Ae(\varphi) : e(\varphi) \end{aligned}$$

Hence

$$\left.\frac{\mathrm{d}}{\mathrm{d}\varepsilon}\left(Ae(\psi+\varepsilon\varphi) : e(\psi+\varepsilon\varphi)\right)\right|_{\varepsilon=0} = Ae(\varphi) : e(\psi) + Ae(\psi) : e(\varphi),$$

and (ii) yields the desired result.

1.1 The Elasticity PDE

(iv) Because of (i) we know that

$$A(\mathscr{O},v,v) = \int_{\mathscr{O}} Ae(v) : e(v)\,\mathrm{d}x$$
$$\geq 2\mu \int_{\mathscr{O}} e(v) : e(v)\,\mathrm{d}x.$$

Now we apply Korn's second inequality (cf. Theorem A.7) and obtain

$$A(\mathscr{O},v,v) \geq \alpha \|v\|_1^2,$$

where $\alpha := 2\mu c' > 0$ (c' is the positive constant from Korn's inequality).

(v) $l(\mathscr{O},\cdot)$ is evidently linear. Continuity follows from Cauchy-Schwarz inequality (cf. Theorem A.10) and the assumption that $f \in L^2(\mathscr{O};\mathbb{R}^2)$ and $g \in H^1(\mathscr{O};\mathbb{R}^2)$:

$$|l(\mathscr{O},\vartheta)| \leq |(f,\vartheta)_0| + |(g,\vartheta)_{L^2(\Gamma_N)}|$$
$$\leq \|f\|_0 \|\vartheta\|_0 + \|g\|_{L^2(\Gamma_N)} \|\vartheta\|_{L^2(\Gamma_N)}$$
$$\leq (\|f\|_0 + \|g\|_1) C \|\vartheta\|_1$$

The last inequality is due to the continuity of the trace operator (see e.g. [Alt02, Eva02]) with some constant C.

(vi) Use (1.5) to write

$$Ae(u) : e(v) = \sum_{i,j} \frac{1}{2}\mu \left(u_{i,j}v_{i,j} + u_{j,i}v_{i,j} + u_{i,j}v_{j,i} + u_{j,i}v_{j,i}\right) + \lambda u_{i,i}v_{i,i}.$$

Then repeated applications of Cauchy-Schwarz inequality (Thm. A.10) and the definition of $\|\cdot\|_1$ (Notation A.5 on page 123) yield the desired result.

□

The next theorem constitutes the weak formulation of the elasticity PDE which is needed in Chapter 2 to develop a finite element solution method.

Theorem 1.4 (Weak form of the linear elasticity PDE). *Finding a solution u to the linear elasticity PDE (1.7) is formally equivalent to finding a solution u to the equations*

$$A(\mathscr{O},u,\vartheta) = l(\mathscr{O},\vartheta) \text{ for all } \vartheta \in V. \tag{1.11}$$

Proof. Suppose u is a solution to (1.7). If we multiply the first line in (1.7) by an arbitrary $\psi \in V$ and integrate over $\mathscr{O}$, we obtain

$$\int_{\mathscr{O}} -\operatorname{div}(Ae(u)) \cdot \psi \, dx = \int_{\mathscr{O}} f \cdot \psi \, dx.$$

Then, taking Definition 1.1 into account and integrating by parts (Theorem A.8) yields

$$\sum_i \int_{\mathscr{O}} f_i \psi_i \, dx = \sum_{i,j} \int_{\mathscr{O}} -\sigma_{ij,j} \psi_i \, dx$$

$$= \sum_{i,j} \left[\int_{\mathscr{O}} \sigma_{ij} \psi_{i,j} \, dx - \int_{\partial \mathscr{O}} \sigma_{ij} \psi_i n_j \, ds \right]$$

$$= \sum_{i,j} \left[\int_{\mathscr{O}} \sigma_{ij} \psi_{i,j} \, dx - \int_{\Gamma_N} \sigma_{ij} \psi_i n_j \, ds - \int_{\Gamma_0} \sigma_{ij} \psi_i n_j \, ds \right]$$

$$= \sum_{i,j} \left[\int_{\mathscr{O}} \sigma_{i,j} \psi_{i,j} \, dx \right] - \int_{\Gamma_N} \underbrace{(\sigma n)}_{=g} \cdot \psi \, ds$$

$$- \int_{\Gamma_0} \underbrace{(\sigma n)}_{=0} \cdot \psi \, ds$$

$$= \sum_{i,j} \left[\int_{\mathscr{O}} \frac{1}{2}(\sigma_{ij} + \sigma_{ji}) \psi_{i,j} \, dx \right] - \int_{\Gamma_N} g \cdot \psi \, ds$$

$$= \sum_{i,j} \left[\int_{\mathscr{O}} \frac{1}{2} \sigma_{ij} (\psi_{i,j} + \psi_{j,i}) \, dx \right] - \int_{\Gamma_N} g \cdot \psi \, ds$$

$$= \int_{\mathscr{O}} \sigma : e(\psi) \, dx - \int_{\Gamma_N} g \cdot \psi \, ds$$

$$= \int_{\mathscr{O}} Ae(u) : e(\psi) \, dx - \int_{\Gamma_N} g \cdot \psi \, ds.$$

Hence we have $\int_{\mathscr{O}} Ae(u) : e(\psi) \, dx = \int_{\mathscr{O}} f \cdot \psi \, dx + \int_{\Gamma_N} g \cdot \psi \, ds$, i.e.

$$A(\mathscr{O}, u, \psi) = l(\mathscr{O}, \psi) \text{ for all } \psi \in V.$$

Conversely, assume that the variational equations are satisfied. Above computations and integration by parts then show that

$$0 = \int_{\mathscr{O}} Ae(u) : e(\psi) \, dx - \int_{\mathscr{O}} f \cdot \psi \, dx - \int_{\Gamma_N} g \cdot \psi \, ds$$

1.1 The Elasticity PDE

$$= \sum_{i,j} \left[\int_{\mathcal{O}} \sigma_{ij} \psi_{i,j} \, dx \right] - \int_{\mathcal{O}} f \cdot \psi \, dx - \int_{\Gamma_N} g \cdot \psi \, ds$$

$$= \sum_{i,j} \left[\int_{\mathcal{O}} -\sigma_{ij,j} \psi_i \, dx + \int_{\partial \mathcal{O}} \sigma_{ij} \psi_i n_j \, ds \right] - \int_{\mathcal{O}} f \cdot \psi \, dx - \int_{\Gamma_N} g \cdot \psi \, ds$$

$$= \int_{\mathcal{O}} (-\operatorname{div}(Ae(u)) - f) \cdot \psi \, dx + \int_{\Gamma_0} (\sigma n) \cdot \psi \, ds$$

$$+ \int_{\Gamma_N} (\sigma n - g) \cdot \psi \, ds.$$

Taking first ψ with compact support in $\mathcal{O}$ gives the state equation. Then, varying the trace function ψ on Γ_N gives the inhomogeneous Neumann boundary condition for u, and varying ψ on Γ_0 gives the homogeneous Neumann boundary condition for u. □

The next theorem states that the problem of finding the elastic deformation u can also be seen as solving a certain minimization problem. This fact will play an important role in Chapter 3. In more general terms, this theorem can be found e.g. in [Cia88, p. 288].

Theorem 1.5 (Existence of a unique solution and variational formulation). *The problem of finding $u \in V$ which satisfies (1.11) has exactly one solution, which is also the unique minimizer of the problem to find $u \in V$ such that*

$$E(\mathcal{O}, u) = \inf_{v \in V} E(\mathcal{O}, v).$$

Proof. The bilinear form $A(\mathcal{O}, \cdot, \cdot)$ is V-elliptic (cf. Lemma 1.3 (iv)), and continuous (cf. Lemma 1.3 (vi)). Furthermore, the linear form $l(\mathcal{O}, \cdot)$ is continuous by Lemma 1.3 (v). By the Lax-Milgram theorem (see Theorem A.6 on page 124), there thus exists one and only one element $\ell \in V$ such that

$$A(\mathcal{O}, \ell, v) = l(\mathcal{O}, v), \quad \forall v \in V.$$

Hence $u = \ell$ is the unique solution to our problem (1.11).

By Lemma 1.3 (ii), the bilinear form $A(\mathcal{O}, \cdot, \cdot)$ is symmetric. Therefore, we can write

$$E(\mathcal{O}, u + v) = \frac{1}{2} A(\mathcal{O}, u + v, u + v) - l(\mathcal{O}, u + v)$$

$$= \frac{1}{2} A(\mathcal{O}, u, u) - l(\mathcal{O}, u) + [A(\mathcal{O}, u, v) - l(\mathcal{O}, v)]$$

$$+ \frac{1}{2}A(\mathcal{O}, v, v)$$
$$= E(\mathcal{O}, u) + [A(\mathcal{O}, u, v) - l(\mathcal{O}, v)] + \frac{1}{2}A(\mathcal{O}, v, v).$$

Consequently, if $A(\mathcal{O}, u, v) = l(\mathcal{O}, v)$, $\forall v \in V$, then by Lemma 1.3 (iv) we have

$$E(\mathcal{O}, u+v) - E(\mathcal{O}, u) = \frac{1}{2}A(\mathcal{O}, v, v) \geq \frac{\alpha}{2}\|v\|_1^2, \quad \forall v \in V,$$

and therefore
$$E(\mathcal{O}, u+v) \geq E(\mathcal{O}, u), \quad \forall v \in V,$$

i.e. u is a minimizer of $E(\mathcal{O}, \cdot)$. Conversely, let u be a minimizer of $E(\mathcal{O}, \cdot)$ and let v be an arbitrary element of V. With the above computations, the following inequality has to be satisfied for all $r \in \mathbb{R}$:

$$0 \leq E(\mathcal{O}, u+rv) - E(\mathcal{O}, u) = r[A(\mathcal{O}, u, v) - l(\mathcal{O}, v)] + \frac{r^2}{2}A(\mathcal{O}, v, v).$$

If $A(\mathcal{O}, u, v) - l(\mathcal{O}, v)$ were $\neq 0$, $r \in \mathbb{R}$ could be chosen such that $E(\mathcal{O}, u+rv) - E(\mathcal{O}, u)$ would be < 0. But this would contradict the assumption that u is a minimizer of $E(\mathcal{O}, \cdot)$. Hence the above inequality implies $A(\mathcal{O}, u, v) = l(\mathcal{O}, v)$, and thus u satisfies (1.11). □

The next Lemma will be useful later to formulate the shape optimization problem as it gives a possibility to treat the requirement that u solves the elasticity PDE as an equality constraint.

Lemma 1.6. *$u \in V$ satisfies (1.11) if and only if*

$$dE(\mathcal{O}, u; \psi) = 0, \quad \forall \psi \in V, \tag{1.12}$$

where $dE(\mathcal{O}, u; \psi) = \frac{d}{d\varepsilon}\left(E(\mathcal{O}, u+\varepsilon\psi)\right)\big|_{\varepsilon=0}$ denotes the first variation of E.

Proof. By the definition of E (see (1.8)) and Lemma 1.3 (iii) we get

$$dE(\mathcal{O}, u; \psi) = \frac{d}{d\varepsilon}\left(\frac{1}{2}A(\mathcal{O}, u+\varepsilon\psi, u+\varepsilon\psi) - l(\mathcal{O}, u+\varepsilon\psi)\right)\bigg|_{\varepsilon=0}$$
$$= A(\mathcal{O}, u, \psi) - l(\mathcal{O}, \psi).$$

Therefore, $dE(\mathcal{O}, u; \psi) = 0$ holds for all $\psi \in V$ if and only if (1.11) holds for u. □

Remark 1.7. *The coercivity of A (cf. Lemma 1.3 (i)) yields that $E(\mathcal{O}, \cdot)$ is strictly convex (see [ET76, Remark 1.1, p. 36] for details), and hence (1.12) are necessary and sufficient optimality conditions for the minimization problem in Theorem 1.5.*

1.2 Shape Optimization Problems

In this section, we give a brief introduction of deterministic shape optimization. The question of how to use material in an efficient way, and design structures and mechanical elements is a very important one in the aerospace industry and the automotive industry, for example. In [BS03], they have among others the design of a lightweight city bus as an application of topology and shape optimization. Another industrial application is the design of suspension triangles in cars, which can be found for example in [AJ05]. Due to applications in physics and engineering the field of shape optimization received a lot of attention in recent years.

The goal in shape optimization is to find a shape among the set of all admissible shapes that minimizes a given objective function. As such it can be seen as a classical optimization problem where one would like to find a feasible point, i.e. a point that satisfies all constraints, which minimizes a certain cost function. However, in shape optimization the competing objects are *shapes*, i.e. subsets of $\mathbb{R}^n$ instead of functions or points. This poses additional difficulties, both theoretically and numerically. Often, the existence of optimal solutions to shape optimization problems cannot be guaranteed, or is even lacking altogether. Then one can try to introduce suitable relaxed formulations, see for example [All02, BB05]. In some special cases, when the objective functional has a specific form or additional geometrical constraints are imposed on the class of admissible domains, existence can be shown. In this work, we focus on the computational aspect, in particular using a level set method.

A classical example of a shape optimization problem which has a solution is the *isoperimetric problem*: the goal is to find an open set $\mathcal{O}$ which maximizes the volume among all open sets with a given and fixed perimeter (cf. [BB05, pp. 3-11]). Minimizing the volume among all open sets with fixed perimeter however is an example for a shape optimization problem that does not have an optimal solution. It is easy to construct a sequence of shapes which all have a given perimeter and whose volumes tend to zero — but zero can never be achieved among the admissible shapes.

The general form of shape optimization problems reads as

$$\min\{J(\mathcal{O}) : \mathcal{O} \in \mathcal{U}_{\text{ad}}\},$$

where $\mathcal{U}_{\text{ad}}$ is the set of admissible domains, and J a cost function which has to be minimized over $\mathcal{U}_{\text{ad}}$. A very important observation is that the set $\mathcal{U}_{\text{ad}}$ — as a set containing subsets of $\mathbb{R}^n$ — has no linear or convex structure. Therefore, it does not make sense to speak of convex functionals for instance in this context.

Shape optimization problems can also be interpreted as *optimal control problems*. They model the behavior of systems that can be *controlled*, i.e. modified, by actions of an operator. The operator cannot influence the state of the system directly, but indirectly by choosing appropriate actions which result in a change of state as a consequence. Accordingly, the two types of variables that are involved in optimal control problems are called *control variables* and *state variables*. The former are the variables the operator is allowed to modify directly to achieve a desired state of the system, in other words a configuration of the state variables. Of course, what the desired state is, depends on the operator's goal. Mathematically, this is usually formulated as a minimization of a *cost functional* which depends on both, the chosen control and the corresponding state of the system. The relation that links the control variables to the state variables is usually called *state equation*. This can be for example an ordinary or partial differential equation.

A typical everyday example of an optimal control problem is driving a car. The driver can only control the accelerator, the brakes, and the steering wheel directly. The speed and the position of the car correspond to the state variables and are a result of the driver's choices. The state equation consists of the mechanical and physical laws that relate the driver's choices to the car's velocity and position. A possible goal for the driver is to get from one position to another as fast as possible. In that case, the objective function of the corresponding control problem would be to minimize the required time. Or, the driver might want to minimize the total gas consumption and is looking for a driving strategy to achieve this. For details on optimal control problems with partial differential equations in general we refer to Tröltzsch [Trö05]. More on the relation to shape optimization problems specifically can be found for example in [BB05, SZ92].

Using the notions of optimal control problems for our shape optimization problems, we can identify the shapes i.e., the domains, with the control variables. Once the shape $\mathcal{O}$ is determined and the corresponding elastic body is subjected to the given forces, its deformation u can be computed by solving the elasticity PDE (1.7). In other words, the deformation u plays the role of the state variables, and the elasticity PDE (1.7) can be identified with the state equation. We can control the deformation u indirectly by choosing a shape $\mathcal{O}$ appropriately. By Lemma 1.6 in Section 1.1 we can treat the state equation as an equality constraint of the shape optimization problem. The advantage of this approach is that one can introduce a *Lagrangian functional* with a *dual variable* which turns out to be very useful to obtain necessary optimality conditions provided all involved functions are differentiable. This dual variable is called *adjoint state* and is the solution of the so-called *adjoint equation* — which turns out to be another elasticity PDE. This formulation will also play an important role in Section 3.1 to develop the analogy

1.2 Shape Optimization Problems

to two-stage stochastic programming. Another technical advantage of the control problem approach is that the differentiation of the state with respect to the control can be avoided. We refer to Section 3.1 for the technical details and constructions.

As in optimal control problems, the goal one would like to achieve is modelled as minimizing a cost functional. A very common choice for an objective functional is the *compliance* which is roughly speaking the inverse of stiffness. So when we are looking for a shape which minimizes the compliance, we seek a shape that is as stiff as possible under the given loadings. Another possibility is to minimize the least square error compared to a given target displacement u_0. We summarize the objective functions we used in the following definition.

Definition 1.8 (Objective functionals). *In general we consider objective functions of the form for $\mathcal{O} \in \mathcal{U}_{ad}$:*

$$\mathbf{J}(\mathcal{O}) = J(\mathcal{O}, u(\mathcal{O})) = \int_{\mathcal{O}} j(x, u(x))\, dx + \int_{\partial \mathcal{O}} k(x, u(x))\, ds. \tag{1.13}$$

The following special cases which match this general form are used for the computational results:

1. *The compliance*

$$\mathbf{J}_1(\mathcal{O}) = \int_{\mathcal{O}} f \cdot u\, dx + \int_{\Gamma_N} g \cdot u\, ds, \tag{1.14}$$

and

2. *the least square error compared to a target displacement u_0*

$$\mathbf{J}_2(\mathcal{O}) = \frac{1}{2} \int_{\mathcal{O}} |u - u_0|^2\, dx. \tag{1.15}$$

In all cases, $u = u(\mathcal{O})$ denotes the solution to the elasticity PDE (1.7).

Remark 1.9. *Note that by the proof of Theorem 1.4, the compliance functional $\mathbf{J}_1$ can also be written as*

$$\mathbf{J}_1 = \int_{\mathcal{O}} Ae(u) : e(u)\, dx.$$

Hence $\mathbf{J}_1$ can be interpreted as the elastic energy stored in the body $\mathcal{O}$ due to the deformation.

As mentioned earlier, minimizing a body's compliance means maximizing its stiffness. This means that in this case optimal shapes tend to use as much material as possible — in particular there would not be any holes in the optimal solution.

However, in many practical applications one can imagine that not only a criterion like compliance should be minimized, but also the amount of used material (especially if there are some costs associated with the material). Therefore it makes sense to penalize the volume of the structure in our shape optimization problem by introducing a nonnegative control parameter $\alpha \in \mathbb{R}, \alpha \geq 0$ and modifying the objective function (1.13) to read

$$\mathbf{J}(\mathcal{O}) = \int_{\mathcal{O}} j(x,u(x))\,dx + \int_{\partial \mathcal{O}} k(x,u(x))\,ds + \alpha \int_{\mathcal{O}} 1\,dx.$$

It is common practice in shape optimization to solve the elasticity problem on the whole working domain D by assuming that $D \setminus \mathcal{O}$ is not really empty but filled with a so-called *ersatz-material* with very small (but positive) values of the Lamé coefficients λ and μ. This approach can be rigorously justified in some cases as shown by Allaire [All02] for instance. It is also the basis for the SIMP method described by Bendsøe and Sigmund in [BS03]. There are some results in this case regarding existence of optimal solutions under additional geometrical, smoothness, or topological constraints (e.g. a perimeter constraint). Details on the question of existence can be found in [All02, AB93, BB05, Buc05, But98, BDM91, BDM93, Cha03, SZ92, Š93].

We emphasize that existence is not an issue in this work, and that we solve the elasticity problem only in the physical domain $\mathcal{O}$. This approach is closer to physical reality but lacks, to our knowledge, any theoretical result concerning existence of solutions. Consequently, we need to solve elasticity PDEs on badly-shaped domains. The solution technique we employed here is described in Chapter 2. Although we are not concerned about existence of optimal shapes here, we also add a perimeter constraint to our problem formulation by introducing another nonnegative control parameter $\beta \in \mathbb{R}, \beta \geq 0$ such that we can finally formulate the shape optimization model we deal with in the following definition.

Definition 1.10 (Shape optimization model). *The shape optimization problem we are concerned with reads as*

$$\min\{\mathbf{J}(\mathcal{O}) : \mathcal{O} \in \mathcal{U}_{ad}, \quad dE(\mathcal{O}, u(\mathcal{O}); \psi) = 0, \ \forall \psi \in V\}, \tag{1.16}$$

with

$$\begin{aligned}\mathbf{J}(\mathcal{O}) = J(\mathcal{O},u(\mathcal{O})) &= \int_{\mathcal{O}} j(x,u(x))\,dx + \int_{\partial \mathcal{O}} k(x,u(x))\,ds \\ &\quad + \alpha \int_{\mathcal{O}} 1\,dx + \beta \int_{\partial \mathcal{O}} 1\,ds,\end{aligned} \tag{1.17}$$

and $\mathcal{U}_{ad} = \{\mathcal{O} \subseteq D : \mathcal{O} \text{ has Lipschitz boundary and } \partial\mathcal{O} \text{ satisfies (1.6)}\}$.

Remark 1.11. *Note that in our computational experiments, the choice $\beta > 0$ did not yield meaningful results. If β was chosen too big, any prescribed hole simply disappeared during the optimization process resulting in a solid structure, whereas a small (but still positive) β did not show any significant effect compared to the choice $\beta = 0$.*

The model (1.16) in Definition 1.10 is the basis for the stochastic extension considered in Chapter 3, where the forces f and g are considered random.

1.3 Two-Stage Stochastic Programming

In this section we give a short introduction to the basic concepts of finite-dimensional stochastic linear optimization. These basic ideas are later applied to our stochastic shape optimization problem (cf. Chapter 3). Note that the notation used here might occasionally clash with notations used in previous sections concerning elasticity theory and PDEs, but in this entire section we are considering finite dimensional linear programs and the notation should therefore be clear from the context.

At the beginning of this chapter we have already mentioned that ignoring uncertainty may lead to inadequate or even plainly wrong decisions. Uncertainty therefore is a very important ingredient in many decision problems that often appear in everyday situations, such as production planning, yacht racing (cf. [PM02]), ground water pollution control, etc., just to mention a few. We refer to the books by Wallace and Ziemba [WZ05], and Birge and Louveaux [BL97] for more applications and details.

Two-stage stochastic programming theory deals with linear programs such as

$$\min\{c^T x : \quad Ax = b, x \geq 0\}, \tag{1.18}$$

with a matrix $A \in \mathbb{R}^{m \times n}$ and vectors $c \in \mathbb{R}^n, b \in \mathbb{R}^m$, where some ingredients are considered uncertain. Those uncertain parameters might be the objective coefficients c[7], the right-hand side vector of the constraints b[8], or the constraint matrix A[9]. Uncertain parameters are represented by random variables where the outcomes of random experiments are denoted by ω. If for example the vector b in (1.18) was random, we would indicate this by writing $b(\omega)$ instead. We denote the set

---

[7] transportation costs usually depend on gas prices, which are random
[8] imagine a production company which has to ensure sufficient availability of their products without knowing future demands
[9] for example crop shortfalls due to uncertain weather conditions

of all outcomes by Ω. The outcomes can be grouped into subsets of Ω called *events*. Events, which one can ask a probability for, are members of a σ-algebra $\mathscr{A}$. Finally, we have a probability measure $\mathbb{P}$ which associates a value $\mathbb{P}(A)$ called *probability* to each event $A \in \mathscr{A}$. The triplet $(\Omega, \mathscr{A}, \mathbb{P})$ is called probability space (cf. [BN95]).

We assume throughout this work that Ω is a finite set $\{\omega_1, \ldots, \omega_S\}$ consisting of a few *states of nature* or *scenarios* with corresponding probabilities $\pi_i = \mathbb{P}(\{\omega_i\})$, for $i = 1, \ldots, S$. In such situations, the knowledge of possible outcomes has to be obtained through some experts' experiences and judgments.

In two-stage stochastic programming we assume that some decisions have to be taken before the uncertain ingredients in the linear program are known. Then, after some time passed by, the values the various random variables actually take can be observed, and we are allowed to make a second decision or take recourse actions. This suggests a chronological order for the decisions that can be made, and also a division of decision variables into two different categories. Those groups of variables gave rise to the term *two-stage* stochastic programming, and are intuitively given as follows:

- Those decisions that have to be taken before the random variables take their actual values are called *first-stage decisions*. Accordingly, the time period when these decisions have to be made is called *first stage*.

- Decisions which can be taken after all realizations of the occurring random variables are known are called *second-stage decisions* — these are determined in the so-called *second stage*.

In the relevant literature such as [BL97, KW94, WZ05] for instance, first-stage variables are always denoted by x and second-stage variables by y. It is also common to write $y(\omega)$ or $y(x, \omega)$ to emphasize that the recourse decisions depend on both, the outcome of the random experiment as well as the decision that has been made in the first stage. This sequence of events and decisions yields the information constraint

$$\text{decide } x \longrightarrow \text{observe } \omega \longrightarrow \text{decide } y = y(x, \omega). \tag{1.19}$$

Note that there are also models where realizations of only a few random variables become known sequentially and the decision maker can make decisions accordingly, taking the newly revealed information and everything known up to that point in time into account. Those models are called *multistage stochastic programs*. They are not a topic in this study and we only refer to [BL97, RS01, RS03b] for the sake of completeness.

1.3 Two-Stage Stochastic Programming

We restrict ourselves to a stochastic two-stage model of the following form:

$$\min\{c^T x + q^T y : Tx + Wy = z(\omega), x \in X, y \in Y\}, \tag{1.20}$$

with finite dimensional polyhedra $X \subseteq \mathbb{R}^n$ and $Y \subseteq \mathbb{R}^m$ together with the information constraint (1.19). The remaining ingredients in (1.20) are the *first stage costs* $c \in \mathbb{R}^n$, the *second stage costs* $q \in \mathbb{R}^m$, the *technology matrix* $T \in \mathbb{R}^{l \times n}$, the *recourse matrix* $W \in \mathbb{R}^{l \times m}$, and a (discrete) random variable $z \colon \Omega \to \mathbb{R}^l$, where $n, m,$ and l are nonnegative integers. Note that in the model (1.20) only the right-hand side vector z of the constraints is uncertain. With W being deterministic, the two-stage program (1.20) is said to have *fixed recourse*. In some applications, there are additionally integer requirements some or all components of x or y have to comply with. We are not concerned with any integer constraints here but refer the reader to [LS03, Sch03a, Sch03b, ST06, Tie05] for completeness.

When looking at the formulation (1.20), it immediately becomes apparent that this very formulation does actually not make much sense, at least if one tries to find a solution. The reason for this is of course the uncertain parameter $z(\omega)$: different realizations of the random variable yield different optimal solutions. How to deal with this issue becomes clear when we rewrite (1.20) into a formulation that emphasizes the two stages in the model. This reformulation reads as

$$\min_x \left\{ c^T x + \min_y \{q^T y : Wy = z(\omega) - Tx, y \in Y\} : x \in X \right\}$$
$$= \min\{c^T x + \Phi(z(\omega) - Tx) : x \in X\}, \tag{1.21}$$

where $\Phi(v) := \min\{q^T y : Wy = v, y \in Y\}$ is the value function of a linear program with parameters on the right-hand side. The cost functional in (1.21) can be interpreted as a function depending on the first-stage decision x and the random outcome ω, i.e., we seek to minimize the cost functional $G(x, \omega) := c^T x + \Phi(z(\omega) - Tx)$. Hence representation (1.21) gives rise to understanding the search for a "best" nonanticipative decision x in the initial random optimization problem (1.20) as the search for a "minimal" member in the family of random variables $\{G(x, \omega) : x \in X\}$ where x is seen as an index varying in the set X. Now we can come back to the problem mentioned earlier that solving (1.20) does not make any sense as long as there is unknown data. Looking for a "minimal" member in the above family of random variables is possible if they are ranked by some relevant criterion. The criterion which is used obviously depends on the particular application.

1.3.1 Expected Value

In the risk-neutral setting, the random variables $G(x, \omega)$ are ranked by their expectations, which results in the (nonlinear) optimization problem

$$\min \{ \mathcal{Q}_{\mathbb{E}}(x) := \mathbb{E}_{\omega}(G(x,\omega)) : x \in X \}. \tag{1.22}$$

The function $\mathcal{Q}_{\mathbb{E}}(x)$ is the expected second-stage value function, and the model (1.22) is the classical two-stage stochastic linear program with fixed recourse originated by Dantzig [Dan55] and Beale [Bea55].

Suppose that the polyhedral sets X and Y in the above formulations are given as $X = \{x \in \mathbb{R}^n : Ax = b, x \geq 0\}$, where A is a matrix in $\mathbb{R}^{\bar{l} \times n}$, and b a vector of matching size, and $Y = \{y \in \mathbb{R}^m : y \geq 0\}^{10}$. Since we assumed that there are finitely many scenarios and probabilities given, problem (1.22) can be equivalently written as

$$\min \{ c^T x + \sum_{i=1}^{S} \pi_i q^T y^i : \begin{array}{l} Ax = b, \\ Tx + Wy^i = z(\omega_i), \ \forall i = 1, \ldots, S, \\ x \in \mathbb{R}^n, x \geq 0, \\ y^i \in \mathbb{R}^m, y^i \geq 0, \ \forall i = 1, \ldots, S \end{array} \}. \tag{1.23}$$

Formulation (1.23) does not contain any unknown ingredients any more: instead of the random variable $z(\omega)$, only its finitely many realizations, i.e. as many as there are scenarios, appear in this model. In other words, (1.23) is simply a linear program, whose number of constraints and variables, however, increases significantly with the number of scenarios. Because of its deterministic nature, problem (1.23) is called *deterministic equivalent program*. The fact that it is a linear program makes it possible to solve it using any available linear programming solver, such as GLPK [Mak], CPLEX [ILO], or Xpress-MP [Das], just to mention a few. In many practical applications however, there might be quite a lot of scenarios and the solution of (1.23) is not efficient at all any more. In that case, one can employ *decomposition algorithms* (see e.g. [BL97, KW94, Pré95, Rus99, VSW69]) to reduce the computational effort by a great amount. These algorithms typically

---

[10]Note that this formulation does not pose a restriction. It would have also been possible to include the constraints $Wy + Tx = z(\omega)$ in the definition of Y. Consequently, Y would have depended on x and ω in this case.

1.3 Two-Stage Stochastic Programming

exploit the special structure of problem (1.23) which can be observed when looking closely at the constraint matrix of problem (1.23): the underlying matrix is of dimension $(\bar{l}+lS) \times (n+mS)$[11], and has the following block structure:

$$\begin{pmatrix} A & & & & & \\ T & W & & & & \\ T & & W & & & \\ T & & & W & & \\ \vdots & & & & \ddots & \\ T & & & & & W \end{pmatrix}. \qquad (1.24)$$

We only consider stochastic programming models with *complete recourse*, as this is the case for our stochastic shape optimization counterpart. This ensures that no first-stage decision and no random outcome can produce infeasible results in the sense that there is no feasible second-stage decision possible any more. Formally this means that $\text{pos}\,W = \mathbb{R}^l$, where $\text{pos}\,W = \{t \in \mathbb{R}^l : \exists y \geq 0 : Wy = t\}$ is the *positive cone* spanned by the columns of W. We assume additionally that the set $\{u \in \mathbb{R}^l : W^T u \leq q\}$ is nonempty and compact, which together with the complete recourse requirement ensures that $\Phi(v)$ is a finite and real number for all $v \in \mathbb{R}^l$. This can be seen by making use of the linear programming duality theorem A.11 on page 125:

$$\Phi(v) = \min\{q^T y : Wy = v, y \geq 0\} = \max\{v^T u : W^T u \leq q\}. \qquad (1.25)$$

Since we assumed that the dual polyhedron is compact and nonempty we know that the maximization problem in (1.25) admits a solution for all $v \in \mathbb{R}^l$. In particular, the compactness implies that the dual polyhedron is a *convex polytope*, i.e., the convex hull of a finite set of points[12]. From [Zie98, p. 52] we know that this polytope is the convex hull of its vertices, and from linear programming theory [Bor01, Chv83, Sch98] we thus know that there is always a vertex in the set of optimal solutions to the maximization problem in (1.25). This is an important observation which allows us to rewrite (1.22) as

$$\min\left\{c^T x + \sum_{\sigma=1}^{S} \pi_\sigma \max_{l=1,\ldots,L} d_l^T (z(\omega_\sigma) - Tx) : x \in X\right\}, \qquad (1.26)$$

where $d_l, l = 1, \ldots, L$ are the vertices of the dual polyhedron $\{u \in \mathbb{R}^l : W^T u \leq q\}$.

---
[11] Note how the dimension strongly depends on the number of scenarios S.
[12] This is a direct consequence from Motzkin's decomposition theorem for polyhedra which can be found for instance in [Sch98, p. 88].

The observations from the previous paragraph show that minimizing $\mathscr{Q}_\mathbb{E}$ amounts to minimizing a piecewise linear convex function over a polyhedron. From an algorithmic point of view, this is advantageous because computing $\mathscr{Q}_\mathbb{E}(x)$ for a given x in (1.22) — i.e. the evaluation of the objective function for only one given point — would require the solution of the linear programs $\min\{q^T y : Wy = z(\omega_\sigma) - Tx, y \geq 0\}$ for all scenarios $\omega_\sigma, \sigma = 1, \ldots, S$. This is not necessary in (1.26) thanks to dual information.

In Chapter 3 we will formulate a two-stage stochastic (infinite dimensional) shape optimization problem as a counterpart to (1.22). The variational formulation of elasticity as described in Theorem 1.5 on page 11 will play the role of the second-stage problem and as such provide an inner minimization problem, analogously to (1.21). The concept of duality will also play an important role and provide information for the shape derivative which will be needed for the descent algorithm. This is worked out in Section 3.1.

1.3.2 Risk Measures

We have seen earlier that solving (1.21) amounts to seeking a "best" member in a family of random variables $\{G(x, \omega) : x \in X\}$. In Section 1.3.1 we simply ranked these random variables by their expectation values. This might not be the desired approach in many practical applications as the expectation simply averages the outcome of the random variables, of which many will occur. In particular, there might be realizations that are rather unlikely but their actual occurrence would have catastrophic consequences. Because of the low probability, however, they would not reflect on the expectation value. In such situations, the matter of *risk aversion* becomes an important issue. Of course, the actual definition of risk depends on the decision maker and is rather arbitrary, which is why we only focus on two risk measures here that have been studied in the context of two-stage stochastic programming, namely *Expected Excess* and *Excess Probability*. There are other risk measures such as the *Value-at-Risk* or the *Conditional Value-at-Risk* commonly used in mathematical finance, and we only refer the reader to [Sch05, ST06, Tie05] for details on those. We introduce the two risk measures we are concerned with in this work in the following definition.

Definition 1.12. *Let $\eta \in \mathbb{R}$ be a preselected cost threshold. Then we can define the following risk measures:*

(i) Expected Excess, *i.e. "the expectation of costs exceeding η":*

$$\mathscr{Q}_{EE_\eta}(G(x,\omega)) := \mathbb{E}\left(\max\{G(x,\omega) - \eta, 0\}\right),$$

1.3 Two-Stage Stochastic Programming

(ii) Excess Probability, i.e. "the probability that costs exceed η":

$$\mathscr{Q}_{EP_\eta}(G(x,\omega)) := \mathbb{P}(\{\omega \in \Omega : G(x,\omega) > \eta\}).$$

In the relevant literature such as [Sch05, ST06, Tie05], they usually study so-called *mean-risk models*

$$\min_{x \in X} \left[\mathbb{E}\{G(x,\omega)\} + \rho \mathscr{R}\{G(x,\omega)\}\right],$$

where $\mathscr{R} \colon \mathscr{Y} \to \mathbb{R}$ denotes a risk measure, $\mathscr{Y}$ the space of all real random cost variables, and $\rho > 0$ a positive weight factor. From a more general point of view, this can be seen as a scalarization of a multiobjective optimization problem which aims to minimize both, the expectation value as well as the risk measure simultaneously (cf. Ehrgott [Ehr05]). However, we shall only deal with the pure risk model later in Sections 3.3 and 3.4 in this study.

Schultz and Tiedemann showed in [ST06, Lemma 4.4] that it is possible to rewrite the expected excess $\mathscr{Q}_{EE_\eta}$ in the form of $\mathscr{Q}_\mathbb{E}$ with a suitably adapted second-stage program which satisfies the required assumptions. However, their models additionally include integer requirements on some of the variables whereas ours do not. Yet it is easy to adapt the quoted Lemma to our situation, and it will turn out to be useful later in Section 3.3 by providing a way to solve the expected excess shape optimization problem numerically. The proof is based on the one presented in [ST06].

Lemma 1.13. *The expected excess $\mathscr{Q}_{EE_\eta}$ can be expressed in the form of $\mathscr{Q}_\mathbb{E}$, namely $\mathscr{Q}_{EE_\eta}(G(x,\omega)) = \mathscr{Q}_\mathbb{E}(\tilde{G}(x,\omega))$ with*

$$\tilde{G}(x,\omega) = \tilde{c}^T \tilde{x} + \tilde{\Phi}(\tilde{z}(\omega) - \tilde{T}\tilde{x}),$$
$$\tilde{\Phi}(\tilde{t}) = \min\{\tilde{q}^T \tilde{y} : \tilde{W}\tilde{y} = \tilde{t}, \tilde{y} \in \mathbb{R}^{\tilde{m}}, \tilde{y} \geq 0\}.$$

Proof. For $\tilde{t} \in \mathbb{R}^{l+1}$ define

$$\tilde{\Phi}(\tilde{t}) := \min\left\{\begin{pmatrix}0\\1\\0\end{pmatrix}^T \begin{pmatrix}y\\v\\w\end{pmatrix} : \begin{pmatrix}W & 0 & 0\\q^T & -1 & 1\end{pmatrix}\begin{pmatrix}y\\v\\w\end{pmatrix} = \tilde{t}, v, w \geq 0, y \geq 0\right\}$$

$$= \min\{v : Wy = \tilde{t}_1, y \geq 0, \ q^T y - \tilde{t}_2 \leq v, 0 \leq v\}.$$

Then we have that

$$\tilde{G}(x,\omega) := \begin{pmatrix}0\\0\end{pmatrix}^T \begin{pmatrix}\eta\\x\end{pmatrix} + \tilde{\Phi}\left(\begin{pmatrix}z(\omega)\\0\end{pmatrix} - \begin{pmatrix}0 & T\\-1 & c^T\end{pmatrix}\begin{pmatrix}\eta\\x\end{pmatrix}\right)$$

$$\begin{aligned}
&= \tilde{\Phi}\left(\begin{pmatrix} z(\omega) - Tx \\ \eta - c^T x \end{pmatrix}\right) \\
&= \min\{v : Wy = z(\omega) - Tx, y \geq 0, \quad q^T y + c^T x - \eta \leq v, 0 \leq v\} \\
&= \max\left\{c^T x - \eta + \underbrace{\min\{q^T y : Wy = z(\omega) - Tx, y \geq 0\}}_{=\Phi(z(\omega) - Tx)}, 0\right\} \\
&= \max\{G(x, \omega) - \eta, 0\}.
\end{aligned}$$

Since we restricted ourselves to the finite scenario case, this yields without further ado that $\mathcal{Q}_{\mathrm{EE}_\eta}(G(x,\omega)) = \mathcal{Q}_{\mathbb{E}}(\tilde{G}(x,\omega))$. □

In Section 3.3 we demonstrate how this Lemma enables us to employ a *barrier method* (see [BGLS03, GK02, NW99, Rus06]) to solve the resulting shape optimization problem with expected excess numerically. We will also present another approach which will simply approximate the max-expression in the objective function such that it can be solved by a gradient descent method.

The algorithmic treatment of stochastic two-stage problems with the excess probability objective leads to linear programs with (additional) binary variables, which indicate for each scenario whether the cost threshold η has been exceeded or not. This is for example discussed in [RS03a, Tie05], and has been applied in [HHW05]. In principle, the ideas from the finite dimensional linear case could be transferred to the infinite dimensional setting of the shape optimization problem with excess probability. One could imagine to do a *branch-and-bound*-like procedure (cf. [Wol98]) where one had to solve nonlinear optimization problems in each node. However, the serious drawback in this approach is the fact that these problems lack convexity. There is no guarantee that the obtained solutions are indeed optimal, which makes it difficult — if not impossible — to obtain meaningful bounds. Therefore, we pursue a different strategy which involves the *Heaviside* function and its appropriate approximation. That way, our problem stays in the class of nonlinear optimization problems, and no integer variables are necessary. The details are described in Section 3.4.

2 Solution of the Elasticity PDE

This chapter is concerned with the solution technique of the elasticity PDE (1.7). Its weak formulation (1.11) constitutes a starting point for the development of a finite element method via the Galerkin method (cf. [Bra03]). One could principally employ the standard finite element method which is described for instance in Braess [Bra03]. However, we aim to solve shape optimization problems using a steepest descent method. In particular, this implies that our optimization variable — which means more precisely in our context the underlying domain the PDE (1.7) needs to be solved on — varies from iteration to iteration. Evidently, this might lead to rather complicated structures in the course of optimization, and would require various grids that frequently adapt to the new shape. Roughly speaking, the more complicated the shapes get, the more triangles we need to resolve the boundary, leading to a high number of degrees of freedom — which is linked directly to the size of the system of linear equations that needs to be solved.

As a remedy, Hackbusch and Sauter [HS97a, HS97b] developed a new class of finite element spaces whose dimensions are independent of the number of geometric details of the physical domains, and which allow to resolve shapes with complicated boundaries with only few degrees of freedom. Furthermore, they make it possible to apply multi-grid methods (see [Hac85]) efficiently to PDEs on such domains. They are called *composite finite elements*, in short CFEs, and have been applied successfully in various applications, e.g. in image based computing [LPR$^+$07]. Additionally to domains with complicated boundary, CFEs can be used for problems involving jumping, i.e. discontinuous, coefficients (cf. [War03]).

The principal idea of the CFE construction is to hierarchically adapt the shape of the finite element basis functions to the behavior of the solution. Therefore, the constructions for Neumann-type boundary conditions [HS97a, HS97b] and Dirichlet-type boundary conditions [RSS06] differ. Our presentation in this chapter mostly follows [Sau02], and is adapted to the elasticity PDE we need to solve. Since we restricted ourselves to the two dimensional case, we will not describe the CFE construction based on a multi-grid method. The computational results in Chapter 5 show that simply solving the resulting linear system using a preconditioned cg method still yields results in reasonable time.

Before we describe the construction of composite finite elements, let us briefly state the very basic ideas and principles of (standard) finite element methods. Suppose we are given a weak formulation of a scalar PDE (in the spirit of Theorem 1.4), i.e.

$$a(u,v) = \langle \ell, v \rangle \quad \text{for all } v \in V, \tag{2.1}$$

with a continuous and coercive bilinear form $a(\cdot,\cdot)$ on an *infinite dimensional* Hilbert space V (for example $H^k(\mathscr{O})$ or $H_0^k(\mathscr{O})$), and a continuous linear form[1] ℓ in the dual space $V^\star$. The main idea is now to restrict (2.1) to a *finite dimensional* subspace V_h of V. h denotes a discretization parameter and should imply that letting h tend to 0 yields a solution of the continuous problem (2.1). We will later specify our choice of the subspace V_h. For now, all we need to know is that V_h is finite dimensional, and therefore there exist a suitable $N \in \mathbb{N}$, and a basis $\{b_1,\ldots,b_N\}$ of V_h. We are now seeking a solution $u_h \in V_h$ determined by

$$a(u_h, v) = \langle \ell, v \rangle \quad \text{for all } v \in V_h.$$

Because of linearity and a finite basis, this is equivalent to

$$a(u_h, b_i) = \langle \ell, b_i \rangle, \quad i = 1,\ldots,N. \tag{2.2}$$

Since we are looking for $u_h \in V_h$, we make the ansatz

$$u_h = \sum_{k=1}^N U_k b_k, \tag{2.3}$$

and obtain a system of linear equations by plugging (2.3) in into (2.2):

$$\sum_{k=1}^N a(b_k, b_i) U_k = \langle \ell, b_i \rangle, \quad i = 1,\ldots,N,$$

which can also be written in matrix-vector notation

$$BU = r, \tag{2.4}$$

with the so-called *system matrix*[2] $B \in \mathbb{R}^{N \times N}$ with entries $B_{ik} := a(b_k, b_i)$, and the right-hand side vector $r \in \mathbb{R}^N$ with entries $r_i := \langle \ell, b_i \rangle$.

Numerically, the given domain $\mathscr{O}$ is divided into finitely many subdomains such as triangles or squares in two dimensions, and tetrahedra or cubes in three dimensions. Then, one considers functions that are polynomials on each of those

---
[1] Note that all prerequisites for the Lax-Milgram Theorem A.6 are satisfied.
[2] Also known as *stiffness matrix*.

2 Solution of the Elasticity PDE

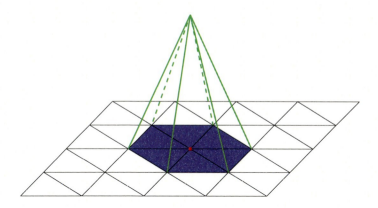

Fig. 2.1: A piecewise linear basis function. It takes the value 1 at the red point, and 0 at all the other grid points. Its support is indicated by the blue triangles.

subdomains. In our case, we use triangles as subdomains, and piecewise linear basis functions. This approach leads to the so-called *Lagrange basis functions*. There are as many basis functions as there are nodes in the grid of triangles, and each of these basis functions takes the value 1 at exactly one node, and 0 at all the other nodes. They are piecewise linear on their support (see Fig. 2.1). For further details on triangulations, the properties they should have, and other finite elements we refer to [Bra03]. We introduce some notation and basic ingredients for our purposes in the following definition.

Definition 2.1. *Let $\mathscr{O} \subseteq \mathbb{R}^2$ be a polygonal domain. We denote a triangulation of $\mathscr{O}$ by $\mathscr{T} := \{\tau_1, \ldots, \tau_M\}$ consisting of open and disjoint triangles, and assume that it is* feasible *or* regular *in the sense of [Bra03, Def. 5.1, p. 58]. Furthermore, we assume that Γ_D and Γ_N are exactly matched by the union of some edges of triangles in $\mathscr{T}$. Then we denote by $V_h := V_h(\mathscr{O})$ the standard finite element space*

$$V_h := \{\varphi \in C^0(\mathscr{O}) : \varphi|_\tau \text{ is linear } \forall \tau \in \mathscr{T}\}. \tag{2.5}$$

Remark 2.2. *The condition in Definition 2.1 that Γ_D and Γ_N need to be exactly matched by the union of some edges of triangles makes it clear that the minimal dimension of the finite element space is directly related to the number of geometric details of the underlying domain. See Fig. 2.2(b) to get an idea how the polygonal domain is obtained.*

2.1 Composite Finite Elements

This section is devoted to the solution of the elasticity PDE (1.7). We start off by recalling[3] its weak formulation (1.11) from page 9: we are looking for $u \in V$ such that
$$A(\mathscr{O}, u, \vartheta) = l(\mathscr{O}, \vartheta) \quad \text{for all } \vartheta \in V.$$

The existence of a unique solution to this problem has already been established in Theorem 1.5. Note that with (1.5), the bilinear form A from (1.9) can also be expressed as

$$A(\mathscr{O}, u, \theta) = \int_{\mathscr{O}} \lambda \operatorname{div}(u) \operatorname{div}(\theta) + 2\mu e(u) : e(\theta) \, dx, \tag{2.6}$$

which is the formulation we will use for the rest of this chapter. The space V contains the Dirichlet boundary condition $u = 0$ on Γ_D. Contrary to the property that a triangulation needs to resolve the domain which we stated in Definition 2.1, we now relax this condition and replace it by the following overlap conditions, imposed on a triangulation $\mathscr{T}$:

$$\mathscr{O} \subseteq \overline{\bigcup_{\tau \in \mathscr{T}} \tau} \text{ and } \forall \tau \in \mathscr{T} : |\tau \cap \mathscr{O}| > 0. \tag{2.7}$$

See Fig. 2.2 for a sketch of a domain $\mathscr{O}$ and the larger domain $\mathscr{O}_{\mathscr{T}} := \operatorname{int} \bigcup_{\tau \in \mathscr{T}} \overline{\tau}$. This figure also indicates the uniform triangular grid we used for all of our numerical computations later. Of course, it would be possible to simply extend the variational formulation (1.11) to the larger domain $\mathscr{O}_{\mathscr{T}}$, but this would lead to a much too large discretization error.

Since the constructions of CFEs differs for Neumann-type and Dirichlet-type boundary conditions, we will consider both cases separately in the next two sections. Later in Section 2.1.3, we will explain how to combine both cases. From now on, we always assume that the domain $\mathscr{O}$ has a polygonal boundary, like the red boundary in Fig. 2.2(b).

2.1.1 Construction for the Neumann Boundary

Let us temporarily pretend that there were no Dirichlet-type boundary conditions. This is of course not really the case and only for presentational purposes, and to explain how to construct CFEs in the case of Neumann-type boundary conditions. Then we can immediately define the *composite finite element space* as follows:

---

[3] Also recall Definition 1.2 on page 6.

2.1 Composite Finite Elements

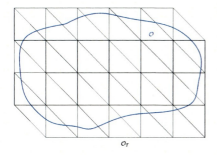

(a) The physical domain is indicated by the blue line. The black grid satisfies the overlap conditions (2.7).

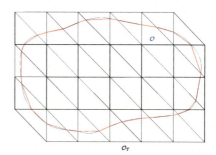

(b) The red lines show the polygonal approximation of the physical domain (blue). This is the domain that is used for the computations.

Fig. 2.2: (a) depicts a physical domain $\mathscr{O}$ and a triangulation satisfying (2.7). (b) demonstrates how a polygonal approximation is obtained.

Definition 2.3. *The composite finite element space V_h^{CFE} for problem* (1.11) *is given by restricting the functions in the standard finite element space $V_h(\mathscr{O}_\mathscr{T})$ (cf.* (2.5) *from Definition 2.1[4]) to the domain $\mathscr{O}$. The finite element discretization to problem* (1.11) *then becomes: Find $u \in V_h^{CFE}$ such that*

$$A(\mathscr{O}, u, \vartheta) = l(\mathscr{O}, \vartheta) \quad \text{for all } \vartheta \in V_h^{CFE}. \tag{2.8}$$

It follows immediately from this definition that CFE basis functions which attain the value 1 far away from the boundary coincide with standard finite element basis functions. Only those CFE basis functions whose support is intersected by the boundary $\partial\mathscr{O}$ are adapted accordingly (cf. Fig. 2.3).

What remains to be explained at this point is how we get from the standard finite element space (2.5) — which was only defined for scalar problems — to the standard finite element space for the vector-valued problem we have at hand here. This is achieved quite easily: Let $N \in \mathbb{N}$ be the number of grid nodes, and $\{b_i\}_{i=1,\ldots,N}$ be the usual scalar Lagrange basis (cf. Fig. 2.1). Then we can define the vector-valued basis functions we need for our two dimensional case by

$$b_i^j := b_i \mathbf{e}_j, \ \forall i = 1,\ldots,N, \ \forall j = 1,2,$$

and the corresponding vector-valued standard finite element space

$$V_h(\mathscr{O}_\mathscr{T}) := \text{span}\left\{b_i^j : i \in \{1,\ldots,N\}, \ j \in \{1,2\}\right\}.$$

---
[4] Note that Definition 2.1 is for scalar problems, but we are dealing with a vector-valued problem now.

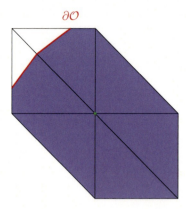

Fig. 2.3: The blue shaded area lies in the inside of the domain $\mathscr{O}$. All six shown triangles would constitute the support for the standard finite element basis function that takes the value 1 at the green colored grid node. The CFE basis function is cut off at the boundary intersection (red line), and its support reduces to the blue area.

While the definition of the composite finite element space seems rather simple, it is not immediately clear how it should be implemented in an efficient way. Those difficulties become evident when taking a closer look at the basis representation of (2.8), following the ideas from standard finite element construction, which we briefly summarized in the beginning of this chapter. A basis in V_h^{CFE} is given by the restrictions

$$b_i^{\text{CFE},j} := b_i^j \big|_{\mathscr{O}}, \ 1 \leq i \leq N, \ 1 \leq j \leq 2,$$

which implies that every function $u \in V_h^{\text{CFE}}$ allows the representation

$$u = \sum_{i=1}^{N} \sum_{j=1}^{2} U_i^j b_i^{\text{CFE},j} \tag{2.9}$$

with coefficient vectors $U^j = \left(U_i^j \right)_{i=1,\ldots,N} \in \mathbb{R}^N$, $j = 1, 2$.

Notation 2.4. *As already indicated in (2.9), we will always use lowercase letters for continuous functions, whereas the corresponding discrete variables will be uppercase and usually denote the nodal values.*

2.1 Composite Finite Elements

The next step is to replace u in (2.8) by its basis representation (2.9) to obtain a system of linear equations $BU = r$ analogously to the standard finite element procedure. The matrix B in that system now has the following block structure:

$$B := \begin{pmatrix} B^{1,1}_{1,1} & \cdots & B^{1,1}_{1,N} & B^{1,2}_{1,1} & \cdots & B^{1,2}_{1,N} \\ \vdots & & \vdots & \vdots & & \vdots \\ B^{1,1}_{N,1} & \cdots & B^{1,1}_{N,N} & B^{1,2}_{N,1} & \cdots & B^{1,2}_{N,N} \\ B^{2,1}_{1,1} & \cdots & B^{2,1}_{1,N} & B^{2,2}_{1,1} & \cdots & B^{2,2}_{1,N} \\ \vdots & & \vdots & \vdots & & \vdots \\ B^{2,1}_{N,1} & \cdots & B^{2,1}_{N,N} & B^{2,2}_{N,1} & \cdots & B^{2,2}_{N,N} \end{pmatrix} = \begin{pmatrix} B^{1,1} & B^{1,2} \\ B^{2,1} & B^{2,2} \end{pmatrix} \in \mathbb{R}^{2N \times 2N}.$$

The individual entries read as

$$B^{l,k}_{i,j} := \int_{\mathcal{O}} \lambda \operatorname{div}\left(b^{\text{CFE},k}_j\right) \operatorname{div}\left(b^{\text{CFE},l}_i\right) + 2\mu e\left(b^{\text{CFE},k}_j\right) : e\left(b^{\text{CFE},l}_i\right) dx,$$

for $1 \leq k,l \leq 2$ and $1 \leq i,j \leq N$. With this, and taking (2.9) and (2.6) into account, we get:

$$A(\mathcal{O}, u, b^{\text{CFE},k}_j) = \sum_{i=1}^{N} \sum_{l=1}^{2} U^l_i \left(\int_{\mathcal{O}} \lambda \operatorname{div}\left(b^{\text{CFE},l}_i\right) \operatorname{div}\left(b^{\text{CFE},k}_j\right) \right.$$
$$\left. + 2\mu e\left(b^{\text{CFE},l}_i\right) : e\left(b^{\text{CFE},k}_j\right) dx \right)$$
$$= \sum_{i=1}^{N} \sum_{l=1}^{2} U^l_i B^{k,l}_{j,i}.$$

Next, we define the right-hand side r as follows:

$$r^k_j := l\left(\mathcal{O}, b^{\text{CFE},k}_j\right) = \int_{\mathcal{O}} f \cdot b^{\text{CFE},k}_j \, dx + \int_{\Gamma_N} g \cdot b^{\text{CFE},k}_j \, ds, \ 1 \leq j \leq N, \ 1 \leq k \leq 2.$$

Then finally we have with

$$U = \left(U^1_1, U^1_2, \ldots, U^1_N, U^2_1, U^2_2, \ldots, U^2_N\right)^T = \left(U^1, U^2\right)^T \in \mathbb{R}^{2N},$$
$$r = \left(r^1_1, r^1_2, \ldots, r^1_N, r^2_1, r^2_2, \ldots, r^2_N\right)^T = \left(r^1, r^2\right)^T \in \mathbb{R}^{2N}$$

that U given by (2.9) satisfies (2.8) if and only if U is a solution of the linear system $BU = r$.

As usual, the matrix B and the vector r are assembled elementwise. This can be seen by making use of the triangulation $\mathcal{T}$ and rewriting $B_{i,j}^{l,k}$ from above as

$$\begin{aligned} B_{i,j}^{l,k} &= \int_{\mathscr{O}} \lambda \operatorname{div}\left(b_j^{\mathrm{CFE},k}\right) \operatorname{div}\left(b_i^{\mathrm{CFE},l}\right) + 2\mu e\left(b_j^{\mathrm{CFE},k}\right) : e\left(b_i^{\mathrm{CFE},l}\right) \mathrm{d}x \\ &= \sum_{\tau \in \mathcal{T}} \int_{\tau \cap \mathscr{O}} \lambda \operatorname{div}\left(b_j^{\mathrm{CFE},k}\right) \operatorname{div}\left(b_i^{\mathrm{CFE},l}\right) + 2\mu e\left(b_j^{\mathrm{CFE},k}\right) : e\left(b_i^{\mathrm{CFE},l}\right) \mathrm{d}x. \end{aligned}$$

When computing $B_{i,j}^{l,k}$ for some indices $i,j \in \{1,\ldots,N\}$, the following observation

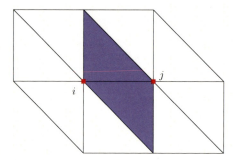

Fig. 2.4: The supports of two nodal basis functions (belonging to nodes i and j, resp.). The colored triangles lie in both supports, and hence only those triangles play a part in the elementwise computation of $B_{i,j}^{l,k}$.

is crucial for an efficient implementation: Only those triangles $\tau \in \mathcal{T}$ need to be considered in the latter sum, which simultaneously belong to both, the support of the nodal basis function associated with node i, and the one associated with node j, as demonstrated in Fig. 2.4. The computation of the matrix is however not done by identifying the appropriate triangles to given indices i and j. Instead — and this is what the term *elementwise* implies — for each triangle $\tau \in \mathcal{T}$, the contribution coming from τ is added to the corresponding entries of the system matrix. This is common practice, and one only needs to compute a local 3×3 matrix per triangle. We will specify this for our case in more detail later in Section 2.1.4. For composite

2.1 Composite Finite Elements

finite elements, the element matrices and vectors are

$$\left.\begin{aligned}
B_{i,j}^{l,k}(\tau) &:= \int_{\tau \cap \mathcal{O}} \lambda \operatorname{div}\left(b_j^{\text{CFE},k}\right) \operatorname{div}\left(b_i^{\text{CFE},l}\right) + 2\mu e\left(b_j^{\text{CFE},k}\right) : e\left(b_i^{\text{CFE},l}\right) \mathrm{d}x \\
f_i^l(\tau) &:= \int_{\tau \cap \mathcal{O}} f \cdot b_i^{\text{CFE},l} \, \mathrm{d}x \\
g_i^l(\tau) &:= \int_{\tau \cap \Gamma_N} g \cdot b_i^{\text{CFE},l} \, \mathrm{d}s,
\end{aligned}\right\} \quad (2.10)$$

for all indices i, j that correspond to vertices in the triangle τ, and for all $1 \leq k, l \leq 2$.

Remark 2.5. *Recall from Definition 2.1 that the triangles are open. As a consequence, it is possible that Γ_N has a nonempty intersection with a triangle's edge. This situation would not be covered by the expression $g_i^l(\tau)$ in (2.10), as $\tau \cap \Gamma_N = \emptyset$ in this case. However, we represent the domain $\mathcal{O}$ by means of a level set function (see Section 4.1 on page 78), and require that its 0-level, i.e. the boundary $\partial \mathcal{O}$, does not pass through any grid node, which prevents the above described situation from occurring.*

(2.10) illustrates the difficulty that arises in the implementation. It is not immediately clear how to compute the integrals over the intersections $\tau \cap \mathcal{O}$ and $\tau \cap \Gamma_N$. Next, we demonstrate how to do this effectively.

2.1.1.1 Implementational Remarks

This paragraph is concerned with deriving quadrature methods which allow to compute the integrals in (2.10) in an efficient way. The idea is to subdivide those triangles that are intersected by the boundary $\partial \mathcal{O}$ and use a composite quadrature rule. Here we assume that each triangle $\tau \in \mathcal{T}$ has the properties that the intersection $\tau \cap \mathcal{O}$ is nonempty and can be subdivided into at most three triangles. In Sauter [Sau02], such triangles are called *simple*. So we already start with a uniform triangulation which consists of only simple triangles. This is the reason why we do not need a refinement procedure as in [Sau02], and why we do not explain the multi-grid technique in this thesis. We denote the set of all subtriangles of a triangle $\tau \in \mathcal{T}$ by $\mathcal{G}(\tau)$. If $\tau \cap \partial \mathcal{O} = \emptyset$, the set $\mathcal{G}(\tau)$ only consists of τ itself, i.e. $\mathcal{G}(\tau) = \{\tau\}$. Otherwise, $\mathcal{G}(\tau)$ consists of one triangle or two triangles, depending on how τ is cut by $\partial \mathcal{O}$ (cf. Fig. 2.5 and Fig. 2.6). In particular, it always holds that $t \cap \mathcal{O} = t$ for all $t \in \mathcal{G}(\tau)$. It is important to note that the subtriangle's vertices are no additional degrees of freedom which is why we also refer to these subtriangles as *virtual triangles*. This construction makes it possible to express an integral over

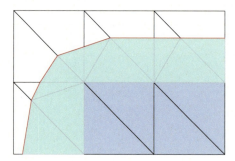

Fig. 2.5: A small part of a domain $\mathcal{O}$ (blue) with its boundary $\partial\mathcal{O}$ (red), and the virtual refined triangles $\mathcal{G}(\tau)$ (green) for some triangles $\tau \in \mathcal{T}$.

$\tau \cap \mathcal{O}$ of some scalar function w as

$$\int_{\tau \cap \mathcal{O}} w(x)\,dx = \sum_{t \in \mathcal{G}(\tau)} \int_{t \cap \mathcal{O}} w(x)\,dx = \sum_{t \in \mathcal{G}(\tau)} \int_{t} w(x)\,dx. \tag{2.11}$$

There are eight ways for $\partial\mathcal{O}$ to intersect a triangle. According to these possibilities, each triangle in the implementation has a certain type. A triangle is assigned type 0 if it is completely inside $\mathcal{O}$, i.e. $\partial\mathcal{O}$ does not intersect this triangle, and type 7 if it is completely outside and as such plays no role in the computation at all. The other possible configurations are depicted in Fig. 2.6. For triangles of types 1 to 6, the vertices with local indices 0 and 1 of the first subtriangle are always on the interface.

We want to employ a hierarchy of basis functions, and express the local basis functions on a triangle $\tau \in \mathcal{T}$ as a combination of the local basis functions living on the subtriangles in $\mathcal{G}(\tau)$. For that we introduce some more notations first.

Notation 2.6. *Let τ be a triangle[5]. We denote its vertices by $\mathbf{P}_{\tau,i}, i \in \{1,2,3\}$. The local linear standard basis functions on τ are denoted by $b_{\tau,i}, i \in \{1,2,3\}$, and are defined as*

$$b_{\tau,i}(\mathbf{P}_{\tau,j}) := \begin{cases} 1 & \text{if } i = j \\ 0 & \text{if } i \neq j, \end{cases} \quad 1 \leq i,j \leq 3. \tag{2.12}$$

---

[5] τ can either be a triangle in $\mathcal{T}$, or a subtriangle of some $\bar{\tau} \in \mathcal{T}$, i.e. $\tau \in \mathcal{G}(\bar{\tau})$.

2.1 Composite Finite Elements

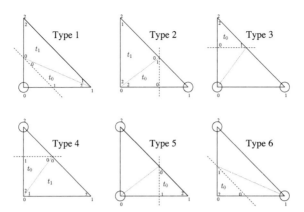

Fig. 2.6: Element types, depending on how the interface (dashed line) intersects the triangle. Vertices on the outside are marked by a circle. The local numbering is shown for the coarse triangles as well as the refined virtual ones. For the differently oriented triangles, the types and numbering is obtained completely analogously by simply rotating the triangles by 180°.

Lemma 2.7. *Let* $\tau \in \mathcal{T}$ *be a triangle.* (2.12) *implies that*

$$b_{\tau,i}|_t = \sum_{j=1}^{3} P_{i,j}^{\tau,t} b_{t,j} \qquad (2.13)$$

holds for any subtriangle $t \subseteq \tau$. *The prolongation matrix* $P^{\tau,t}$ *is given by*

$$P_{i,j}^{\tau,t} := b_{\tau,i}(\mathbf{P}_{t,j}), \ 1 \leq i,j \leq 3.$$

Proof. We have to verify (2.13) for the vertices of the subtriangle t, namely for $\mathbf{P}_{t,k}$ for $k \in \{1, 2, 3\}$. Because of the Lagrange property (2.12) we have

$$b_{t,j}(\mathbf{P}_{t,k}) = \delta_{jk},$$

δ being Kronecker's delta. Hence,

$$\sum_{j=1}^{3} P_{i,j}^{\tau,t} b_{t,j}(\mathbf{P}_{t,k}) = \sum_{j=1}^{3} P_{i,j}^{\tau,t} \delta_{jk}$$

$$= \sum_{j=1}^{3} b_{\tau,i}(\mathbf{P}_{t,j}) \delta_{jk} = b_{\tau,i}(\mathbf{P}_{t,k}).$$

□

Now we use the above results and observations to find a way to compute a local element matrix $B_{i,j}^{l,k}(\tau)$ for some triangle $\tau \in \mathscr{T}$. Recall from (2.10) that the local matrices $B^{l,k}(\tau) := \left(B_{i,j}^{l,k}(\tau)\right)_{ij} \in \mathbb{R}^{3\times 3}$, $1 \leq k,l, \leq 2$, have the representations

$$B_{i,j}^{l,k}(\tau) := \int_{\tau \cap \mathscr{O}} \lambda \operatorname{div}\left(b_j^{\mathrm{CFE},k}\right) \operatorname{div}\left(b_i^{\mathrm{CFE},l}\right) + 2\mu e\left(b_j^{\mathrm{CFE},k}\right) : e\left(b_i^{\mathrm{CFE},l}\right) \, dx.$$

This yields in combination with (2.13) and (2.11) for each $\tau \in \mathscr{T}$:

$$B^{l,k}(\tau) = \sum_{t \in \mathscr{G}(\tau)} P^{\tau,t} B^{l,k}(t) \left(P^{\tau,t}\right)^T. \tag{2.14}$$

This can be seen as follows:

$$B^{l,k}(\tau) = \left(B_{i,j}^{l,k}(\tau)\right)_{ij},$$

$$B_{i,j}^{l,k}(\tau) = \sum_{t \in \mathscr{G}(\tau)} \int_{t \cap \mathscr{O}} \lambda \operatorname{div}(b_{\tau,j}\mathbf{e}_k) \operatorname{div}(b_{\tau,i}\mathbf{e}_l) + 2\mu e(b_{\tau,j}\mathbf{e}_k) : e(b_{\tau,i}\mathbf{e}_l) \, dx$$

(using (2.13) now)

$$= \sum_{t \in \mathscr{G}(\tau)} \int_{t \cap \mathscr{O}} \lambda \operatorname{div}\left(\sum_{\kappa=1}^{3} P_{j,\kappa}^{\tau,t} b_{t,\kappa} \mathbf{e}_k\right) \operatorname{div}\left(\sum_{\nu=1}^{3} P_{i,\nu}^{\tau,t} b_{t,\nu} \mathbf{e}_l\right)$$

$$+ 2\mu e \left(\sum_{\kappa=1}^{3} P_{j,\kappa}^{\tau,t} b_{t,\kappa} \mathbf{e}_k\right) : e\left(\sum_{\nu=1}^{3} P_{i,\nu}^{\tau,t} b_{t,\nu} \mathbf{e}_l\right) dx$$

$$= \sum_{t \in \mathscr{G}(\tau)} \sum_{\kappa=1}^{3} \sum_{\nu=1}^{3} B_{\nu,\kappa}^{l,k}(t) P_{j,\kappa}^{\tau,t} P_{i,\nu}^{\tau,t}$$

$$= \sum_{t \in \mathscr{G}(\tau)} \sum_{\kappa=1}^{3} P_{j,\kappa}^{\tau,t} \left(P^{\tau,t} B^{l,k}(t)\right)_{i\kappa}$$

$$= \sum_{t \in \mathscr{G}(\tau)} \left(P^{\tau,t} B^{l,k}(t) \left(P^{\tau,t}\right)^T\right)_{ij}.$$

Remark 2.8. *We can summarize the procedure to compute the element matrix $B^{l,k}(\tau)$ for a coarse grid triangle $\tau \in \mathscr{T}$ and indices $k,l \in \{1,2\}$ in the following two steps:*

1. *Compute and store the element matrices $B^{l,k}(t)$ for all subtriangles $t \in \mathscr{G}(\tau)$.*

2. *Compute the element matrix $B^{l,k}(\tau)$ according to (2.14).*

2.1 Composite Finite Elements

Note that in case that τ is an inner triangle, i.e. $\tau \cap \mathcal{O} = \tau$ and $\tau \cap \partial \mathcal{O} = \emptyset$, the set $\mathcal{G}(\tau)$ consists of only τ itself. Consequently, the prolongation matrices defined in Lemma 2.7 are simply identity matrices. This means that the two steps described above become trivial but remain valid for inner triangles that are not intersected by the boundary as well.

2.1.2 Construction for the Dirichlet Boundary

So far we have described how to construct composite finite element basis functions for triangles intersected by the Neumann boundary. In doing so in the previous Section 2.1.1, we totally neglected the fact that the elasticity PDE (1.7) additionally contains Dirichlet-type boundary conditions, namely $u = 0$ on Γ_D. In the weak formulation (1.11), this is "hidden" in the function space V (cf. Definition 1.2 on page 6). We demonstrate in this section how to construct composite finite element basis functions in the case of Dirichlet-type boundary conditions. Similarly to the previous section, we are not concerned with the Neumann part of the boundary here, and therefore pretend that there were no Neumann-type boundary conditions. Again, we follow the construction procedure from [Sau02], for further details also see [RSS06]. The constructions of composite finite elements with Dirichlet boundary conditions consists of four steps.

Step 1: Overlapping grid Just as in Section 2.1.1, we assume that we have a finite element grid $\mathcal{T} = \{\tau_1, \ldots, \tau_M\}$ consisting of open and disjoint triangles without hanging nodes, and additionally satisfying the overlap conditions (2.7). We denote the set of all vertices of triangles in $\mathcal{T}$, those are in other words simply the grid nodes, by $\Theta = \{x_i : 1 \leq i \leq N, x_i \text{ is a grid node }\}$. The set of vertices of one triangle τ is denoted by $\text{vert}(\tau)$.

Step 2: Marking the degrees of freedom We distinguish between two types of nodes:

- *free nodes* are those where the degrees of freedom are located. Those nodes are inner nodes and not vertices of triangles that are intersected by the boundary.

- *slave nodes* are those where the function values are constrained in a way such that the Dirichlet boundary conditions are satisfied.

We introduce some more notations which allow us to define the above described types of nodes.

Definition 2.9. *If $x_i \in \Theta$ is a grid node, we can define its* **triangle neighborhood** *$\mathscr{T}_i$ as*

$$\mathscr{T}_i := \{\tau \in \mathscr{T} : x_i \in \overline{\tau}\}.$$

Let

$$\Theta^{\Gamma_D} := \bigcup_{\tau \in \mathscr{T}: |\tau \cap \Gamma_D| > 0} \text{vert}(\tau), \text{ and } \mathscr{T}^{\Gamma_D} := \bigcup_{i \in \Theta^{\Gamma_D}} \mathscr{T}_i.$$

The complements of these sets are $\Theta^{in} := \Theta \setminus \Theta^{\Gamma_D}$ and $\mathscr{T}^{in} := \mathscr{T} \setminus \mathscr{T}^{\Gamma_D}$. Then we can introduce the **inner domain**

$$\mathscr{O}^{in} := \text{int}\left(\bigcup_{\tau \in \mathscr{T}^{in}} \overline{\tau}\right).$$

The degrees of freedom are associated with the nodes in Θ^{in} which constitute the **free nodes**. **Slave nodes** *are those nodes in Θ^{Γ_D}. See also Fig. 2.7 for a sketch of a part of a domain which is intersected by the Dirichlet boundary, and the different types of nodes and triangles.*

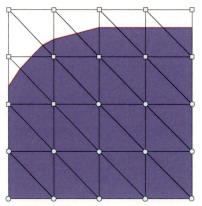

(a) Slave nodes in Θ^{Γ_D} are indicated by squares. Free nodes in Θ^{in} are marked by circles.

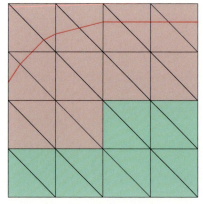

(b) Inner triangles in $\mathscr{T}^{in}$ are shown in green. The brown colored triangles belong to $\mathscr{T}^{\Gamma_D}$.

Fig. 2.7: The physical domain is indicated by the blue shaded triangles. The red line is now part of the Dirichlet boundary Γ_D. (a) shows a small part of $\mathscr{O}$ next to its Dirichlet boundary with marked *slave* and *free* nodes. In (b), the same part of the domain is depicted, showing the triangles belonging to $\mathscr{T}^{\Gamma_D}$ and $\mathscr{T}^{in}$.

2.1 Composite Finite Elements

Step 3: Definition of an extrapolation operator In this step, the supports of basis functions defined on inner triangles $\tau \in \mathcal{T}^{\text{in}}$ are extended to triangles in $\mathcal{T}^{\Gamma_D}$. Function values at slave nodes in Θ^{Γ_D}, which are necessary for the computation, are determined by function values at close by inner nodes in Θ^{in}. To achieve this, we define an extrapolation operator $\mathcal{E}: V_h^{\text{in}} \to V_h$, where V_h is the standard finite element space for $\mathcal{O}_\mathcal{T}$ and the triangulation $\mathcal{T}$ as defined earlier (the vector-valued counterpart of (2.5)), whereas V_h^{in} denotes the standard finite element space for the domain $\mathcal{O}^{\text{in}}$ (without boundary conditions) and the triangulation $\mathcal{T}^{\text{in}}$. Before we come to the definition of $\mathcal{E}$, we need to introduce additional notation.

Notation 2.10.

(i) *For a function $v \in V_h^{\text{in}}$ and a triangle $\tau \in \mathcal{T}$, we denote the analytic extension of $v|_\tau$ to $\mathbb{R}^2$ by $v_\tau^\star: \mathbb{R}^2 \to \mathbb{R}$.*

(ii) *Let $x \in \mathbb{R}^2$. A triangle in $\mathcal{T}^{\text{in}}$ with minimal distance from x is denoted by τ_x.*

(iii) *If $x \in \mathbb{R}^2$, then $x^{\Gamma_D} \in \Gamma_D$ denotes a point on the boundary with minimal distance from x.*

(iv) *$\varsigma > 0$: parameter to control how far away a triangle τ_x is allowed to be for a slave node $x \in \Theta^{\Gamma_D}$, to still have an effect.*

Since the image of $\mathcal{E}v$, for some $v \in V_h^{\text{in}}$, is in the standard finite element space V_h which consists of piecewise linear functions, it suffices to specify the values of $\mathcal{E}v$ at the grid points in Θ. The function $\mathcal{E}v$ is then the unique linear interpolation of these nodal values. Now let $v \in V_h^{\text{in}}$, and define for $x \in \Theta$

$$(\mathcal{E}v)(x) := \begin{cases} v(x) & \text{if } x \in \Theta^{\text{in}}, \\ v_{\tau_x}^\star(x) - v_{\tau_x}^\star(x^{\Gamma_D}) & \text{if } x \in \Theta^{\Gamma_D} \text{ and } \text{dist}(x, \tau_x) \leq \varsigma \, \text{diam}\, \tau_x, \quad (2.15) \\ 0 & \text{otherwise,} \end{cases}$$

where $\text{dist}(x, \tau_x)$ denotes the distance between the point x and the triangle τ_x and $\text{diam}\,\tau_x$ the diameter of the triangle τ_x.

Step 4: Definition of composite finite elements with homogeneous boundary conditions Like in Section 2.1.1 we assume that all triangles in our triangulation $\mathcal{T}$ are simple. As before, the set of all (virtual) subtriangles of a triangle $\tau \in \mathcal{T}$ is $\mathcal{G}(\tau)$. Then we consider the following triangulation

$$\tilde{\mathcal{T}} := \bigcup_{\tau \in \mathcal{T}} \mathcal{G}(\tau), \qquad (2.16)$$

which consists of all triangles in $\mathcal{T}$ and additionally all virtual subtriangles. Next, let $\tilde{V}_h$ be the standard finite element space for the triangulation $\tilde{\mathcal{T}}$. Similarly to the prolongation matrices from Lemma 2.7 in the Neumann-type boundary construction, we employ a *modification operator* $\mathcal{M} : V_h \to \tilde{V}_h$ to define a composite finite element space. This modification operator adapts a standard finite element function $u \in V_h$ in a neighborhod of the boundary to the boundary conditions. Since the mesh $\tilde{\mathcal{T}}$ resolves the domain, the application of the operator $\mathcal{M}\mathcal{E}$ to V_h^{in} results in a finite element space which satisfies the homogeneous boundary conditions. Let $\tilde{\Theta}$ be the nodal points of $\tilde{\mathcal{T}}$. Then $\tilde{\Theta}^{\Gamma_D}$ and $\tilde{\Theta}^{in}$ are defined accordingly to Definition 2.9. Again, just as in the definition of the extrapolation operator $\mathcal{E}$, it suffices to define $\mathcal{M}$ on the grid nodes $\tilde{\Theta}$: For $u \in V_h$ and $x \in \tilde{\Theta}$, define

$$(\mathcal{M}u)(x) := \begin{cases} u(x) - u\left(x^{\Gamma_D}\right) & \text{if } x \in \tilde{\Theta}^{\Gamma_D}, \\ u(x) & \text{if } x \in \tilde{\Theta}^{in}. \end{cases} \tag{2.17}$$

Note that this means that

$$(\mathcal{M}u)(x_i) = \begin{cases} u(x_i) & \text{if } x_i \text{ is a vertex in } \tilde{\mathcal{T}} \text{ and } x_i \notin \Gamma_D, \\ 0 & \text{if } x_i \text{ is a vertex in } \tilde{\mathcal{T}} \text{ and } x_i \in \Gamma_D. \end{cases}$$

In other words, the homogeneous Dirichlet boundary conditions are satisfied on Γ_D. Finally, we can define the composite finite element space.

Definition 2.11. *The composite finite element space for Dirichlet boundary conditions is given by*

$$V_h^{CFE} := \left\{ \mathcal{M}\mathcal{E}u : u \in V_h^{in} \right\}.$$

The algorithmic realization of the composite finite elements for the Dirichlet case is similar to the one in the Neumann case. We take a closer look at it now in the following paragraph.

2.1.2.1 Implementational Remarks

We start with the algorithmic realization of the modification operator $\mathcal{M} : V_h \to \tilde{V}_h$. For that we define for $\tau \in \mathcal{T}$ and $t \in \mathcal{G}(\tau)$ the local 3×3 modification matrices by[6]

$$P_{i,j}^{\tau,t} := \begin{cases} b_{\tau,i}(x_j) & \text{if } x_j \notin \Gamma_D, \\ 0 & \text{if } x_j \in \Gamma_D, \end{cases} \quad \forall x_j \in \text{vert}(t), \forall x_i \in \text{vert}(\tau).$$

---

[6] In analogy to the Neumann case, we also denote the modification matrices by $P^{\tau,t}$.

2.1 Composite Finite Elements

Then we can generate the linear system for the space

$$\{\mathscr{M}u : u \in V_h\} \tag{2.18}$$

by a recursion which is of the same form as (2.14). Note that the space (2.18) is a subspace of V_h^{CFE}: Suppose that $v \in V_h^{\text{CFE}}$. Then there exists a $\bar{u} \in V_h^{\text{in}}$ such that $v = \mathscr{M}\mathscr{E}\bar{u}$. By definition of the extrapolation operator $\mathscr{E}$, $\mathscr{E}\bar{u}$ is in V_h. Therefore, $v \in \{\mathscr{M}u : u \in V_h\}$ which shows that $V_h^{\text{CFE}} \subseteq \{\mathscr{M}u : u \in V_h\}$.

So exactly as in the Neumann case, we can compute the local element matrix for a triangle $\tau \in \mathscr{T}$ as

$$B^{l,k}(\tau) = \sum_{t \in \mathscr{G}(\tau)} P^{\tau,t} B^{l,k}(t) \left(P^{\tau,t}\right)^T, \quad 1 \leq k,l \leq 2. \tag{2.19}$$

This time however, the basis functions are adapted to the homogeneous Dirichlet boundary conditions. In this way, the global system matrices[7] $\tilde{B}^{l,k}$ can be assembled for the space (2.18) with the usual techniques.

The next step is to include the extrapolation operator $\mathscr{E}$ to extend the basis functions from triangles in $\mathscr{T}^{\text{in}}$ to those triangles which are intersected by the boundary, i.e. those in $\mathscr{T}^{\Gamma_D}$. As mentioned earlier, it suffices to define $\mathscr{E}$ at the grid nodes, as the function values are the unique linear interpolation of these nodal values. This will lead to a matrix representation for $\mathscr{E}$ which we denote by $E \in \mathbb{R}^{N \times N^{\text{in}}}$, where N^{in} is the dimension of the space V_h^{in} — in other words, the number of free nodes. The transposed matrix E^T then maps values on slave nodes back to inner nodes. We will specify this matrix below. Using E, we can obtain the system matrix for the composite finite element space V_h^{CFE} as

$$B^{l,k} := E^T \tilde{B}^{l,k} E, \quad 1 \leq k,l \leq 2. \tag{2.20}$$

The extrapolation process increases the supports of basis functions which are close to the boundary. Algorithmically, the support of $\mathscr{E}b_i^{\text{in}}$, where b_i^{in} is the standard finite element basis function for the space V_h^{in} for node $x_i \in \Theta^{\text{in}}$, is computed according to

$$\mathscr{E}b_i^{\text{in}} = \sum_{j \in \mathfrak{P}_i} \left(\mathscr{E}b_i^{\text{in}}\right)(x_j) b_j, \tag{2.21}$$

with the standard finite element basis function b_j for the space V_h for node $x_j \in \Theta$. The set $\mathfrak{P}_i$ for $i \in \Theta^{\text{in}}$ is computed in the following way:

1. For all slave nodes $x \in \Theta^{\Gamma_D}$, find its closest inner triangle $\tau_x \in \mathscr{T}^{\text{in}}$.

---
[7] Recall the construction of the system matrix in Section 2.1.1, in particular its block structure.

2. For any inner triangle $\tau \in \mathscr{T}^{\text{in}}$, generate the set of all indices of slave nodes which have τ as closest inner triangle, i.e.

$$\Theta_\tau := \{j : x_j \in \Theta^{\Gamma_D} \text{ and } \tau = \tau_{x_j}\}.$$

3. For each inner node $x_i \in \Theta^{\text{in}}$ initialize the set $\mathfrak{P}_i := \{i\}$.

4. For any inner triangle $\tau \in \mathscr{T}^{\text{in}}$ and for all vertices $x_i \in \text{vert}(\tau)$, update $\mathfrak{P}_i := \mathfrak{P}_i \cup \Theta_\tau$.

Then representation (2.21) leads to the definition of the entries of the extrapolation matrix:

$$E_{ji} := \begin{cases} (\mathscr{E} b_i^{\text{in}})(x_j) & \text{if } j \in \mathfrak{P}_i, \\ 0 & \text{otherwise,} \end{cases} \quad \text{for } 1 \leq j \leq N, \, 1 \leq i \leq N^{\text{in}}.$$

To check this, we plug in (2.21) into the above definition for E:

$$\begin{aligned} E_{ji} &= \begin{cases} (\mathscr{E} b_i^{\text{in}})(x_j) & \text{if } j \in \mathfrak{P}_i, \\ 0 & \text{otherwise} \end{cases} \\ &= \begin{cases} \sum_{k \in \mathfrak{P}_i} (\mathscr{E} b_i^{\text{in}})(x_k) b_k(x_j) & \text{if } j \in \mathfrak{P}_i, \\ 0 & \text{otherwise.} \end{cases} \end{aligned} \quad (2.22)$$

Now, for $j, k \in \mathfrak{P}_i$ we know from the construction of the set $\mathfrak{P}_i$ that

$$(\mathscr{E} b_i^{\text{in}})(x_k) b_k(x_j) = \begin{cases} b_i^{\text{in}}(x_k) & \text{if } x_k \in \Theta^{\text{in}}, \\ (b_i^{\text{in}})^\star(x_k) - (b_i^{\text{in}})^\star(x_k^{\Gamma_D}) & \text{if } x_k \in \Theta^{\Gamma_D} \text{ and} \\ & \text{dist}(x_k, \tau_{x_k}) \leq \varsigma \, \text{diam}(\tau_{x_k}), \\ 0 & \text{otherwise} \end{cases} b_k(x_j) \quad (2.23)$$

By construction, the set $\mathfrak{P}_i$ contains exactly one inner point in Θ^{in}, namely x_i. With the Lagrange property of the basis functions, we therefore get with (2.22) and (2.23):

$$E_{ji} = \begin{cases} 1 & \text{if } j = i, \\ (b_i^{\text{in}})^\star(x_j) - (b_i^{\text{in}})^\star(x_j^{\Gamma_D}) & \text{if } j \in \mathfrak{P}_i \setminus \{i\} \\ & \text{and } \text{dist}(x_j, \tau_{x_j}) \leq \varsigma \, \text{diam}(\tau_{x_j}), \\ 0 & \text{otherwise,} \end{cases}$$

which indeed coincides with the definition (2.15) of $\mathscr{E}$.

2.1 Composite Finite Elements

Remark 2.12. *Let us summarize the procedure to compute the system matrix $B^{l,k}$ for indices $k, l \in \{1,2\}$:*

1. *Generate the element matrices for $\tilde{\mathcal{T}}$.*
2. *Use these element matrices to obtain the local element matrices for $\mathcal{T}$ via the formula (2.19).*
3. *Using the element matrices from the previous step, assemble the corresponding system matrix for the space (2.18) in the usual way.*
4. *Construct the global extrapolation matrix E in a sparse format.*
5. *Generate the system matrix for V_h^{CFE} according to representation (2.20).*

Note that although our implementation contains the extrapolation operator, we did not use it in most of our computational experiments. This is because our test instances only contain a simple Dirichlet boundary, i.e. a straight line, and we simply assume that the Dirichlet boundary matches edges of triangles exactly. Especially the CPU time needed for step 5 can be saved that way, which already results in significant speed improvements.

We finish this section with a small one dimensional example to demonstrate how the extrapolation and modification operators work.

2.1.2.2 Simple 1D Example

Suppose we have the situation depicted in Fig. 2.8, where only a small part right next to the boundary Γ_D is shown. There we have two slave nodes $x_3, x_4 \in \Theta^{\Gamma_D}$ and two inner nodes $x_1, x_2 \in \Theta^{\text{in}}$. The boundary intersects the "triangle"[8] exactly in the middle at a point x^{Γ_D}. This intersection point is the closest boundary point for both, x_3 as well as x_4. For both slave nodes, the closest inner triangle is one and the same, denoted by τ_x in the picture. According to the definition of the composite finite element space V_h^{CFE} in Definition 2.11, the values of the basis functions $b_{x_1}^{\text{in}}, b_{x_2}^{\text{in}} \in V_h^{\text{in}}$ at the slave nodes x_3, x_4 are $\left(\mathcal{M}\mathcal{E} b_{x_1}^{\text{in}}\right)(x_3)$ and $\left(\mathcal{M}\mathcal{E} b_{x_2}^{\text{in}}\right)(x_4)$, respectively. Using (2.15) and (2.17), and assuming that the distance parameter ς is chosen big enough such that the distance condition in (2.15) is satisfied for both slave nodes, we can compute the values of the extended basis functions in the CFE space as follows (also see Fig. 2.9):

$$\left(\mathcal{E} b_{x_1}^{\text{in}}\right)(x_3) = \left(b_{x_1}^{\star}\right)(x_3) - \left(b_{x_1}^{\star}\right)\left(x^{\Gamma_D}\right) = -1 - (-1.5) = \frac{1}{2},$$

---
[8] Actually, it is a line segment in one dimension.

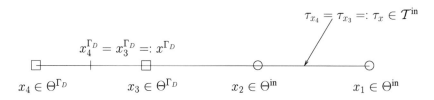

Fig. 2.8: Set up for a small one dimensional example to illustrate how the extrapolation operator works. Slave nodes are marked by squares, free nodes by circles.

$$\left(\mathscr{E}b_{x_1}^{\text{in}}\right)(x_4) = \left(b_{x_1}^{\star}\right)(x_4) - \left(b_{x_1}^{\star}\right)\left(x^{\Gamma_D}\right) = -2 - (-1.5) = -\frac{1}{2},$$

$$\left(\mathscr{E}b_{x_2}^{\text{in}}\right)(x_3) = \left(b_{x_2}^{\star}\right)(x_3) - \left(b_{x_2}^{\star}\right)\left(x^{\Gamma_D}\right) = 2 - 2.5 = -\frac{1}{2},$$

$$\left(\mathscr{E}b_{x_2}^{\text{in}}\right)(x_4) = \left(b_{x_2}^{\star}\right)(x_4) - \left(b_{x_2}^{\star}\right)\left(x^{\Gamma_D}\right) = 3 - 2.5 = \frac{1}{2},$$

leading to

$$\left(\mathscr{M}\mathscr{E}b_{x_1}^{\text{in}}\right)(x_3) = \left(\mathscr{E}b_{x_1}^{\text{in}}\right)(x_3) - \left(\mathscr{E}b_{x_1}^{\text{in}}\right)\left(x^{\Gamma_D}\right) = \frac{1}{2} - 0 = \frac{1}{2},$$

$$\left(\mathscr{M}\mathscr{E}b_{x_1}^{\text{in}}\right)(x_4) = \left(\mathscr{E}b_{x_1}^{\text{in}}\right)(x_4) - \left(\mathscr{E}b_{x_1}^{\text{in}}\right)\left(x^{\Gamma_D}\right) = -\frac{1}{2} - 0 = -\frac{1}{2},$$

$$\left(\mathscr{M}\mathscr{E}b_{x_2}^{\text{in}}\right)(x_3) = \left(\mathscr{E}b_{x_2}^{\text{in}}\right)(x_3) - \left(\mathscr{E}b_{x_2}^{\text{in}}\right)\left(x^{\Gamma_D}\right) = -\frac{1}{2} - 0 = -\frac{1}{2},$$

$$\left(\mathscr{M}\mathscr{E}b_{x_2}^{\text{in}}\right)(x_4) = \left(\mathscr{E}b_{x_2}^{\text{in}}\right)(x_4) - \left(\mathscr{E}b_{x_2}^{\text{in}}\right)\left(x^{\Gamma_D}\right) = \frac{1}{2} - 0 = \frac{1}{2}.$$

This shows that the value of $\left(\mathscr{M}\mathscr{E}b_{x_i}^{\text{in}}\right)(x^{\Gamma_D})$ for $i = 1, 2$ is indeed 0 as the linear interpolation of the above nodal values at exactly the middle point between x_3 and x_4.

2.1.3 Mixed Boundary Conditions

After having described the separate cases for Neumann and Dirichlet boundary conditions in Sections 2.1.1 and 2.1.2, respectively, we can now easily combine these two ways of construction. The only difference is that not the whole boundary is either Γ_N or Γ_D, and the set of slave nodes, which we have already denoted by Θ^{Γ_D}, now only contains nodes next to the portion of the Dirichlet boundary Γ_D. The inner grid arises accordingly, and the construction for the Dirichlet part of the

2.1 Composite Finite Elements

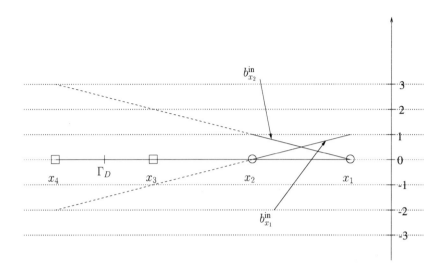

Fig. 2.9: Here the basis functions $b_{x_1}^{in}, b_{x_2}^{in} \in V_h^{in}$ are drawn in solid lines on τ_x (cf. Fig. 2.8). The dashed lines show their extensions to the slave nodes which are needed for the extrapolation operator.

boundary can be carried out literally as described in Section 2.1.2. The Neumann boundary part is not affected at all by the adapted definition of Θ^{Γ_D}, which follows immediately from the definitions of the extrapolation operator (2.15) and the modification operator (2.17). Therefore, the composite finite element functions next to the Dirichlet boundary are modified just as in the pure Dirichlet boundary case, whereas the composite finite element functions in a neighborhood of the Neumann boundary are restrictions of the overlapping triangulation, just as in the pure Neumann boundary case.

In Fig. 2.10(a), we consider a small example of a rectangular elastic body, fixed on its left edge, subjected to a surface load on its right edge, pulling it down. In 2.10(b), one can see how the boundary intersects the overlapping triangulation. This example illustrates the effect of the extrapolation operator as the Dirichlet boundary does not match edges of triangles. The solution u to the PDE (1.7) can be seen in Fig. 2.11(a). Fig. 2.11(b) demonstrates that u is indeed approximately 0 on Γ_D.

Finally, for the sake of completeness, we refer to [HS97b] for approximation properties in the Neumann case, and to [RSS06] for such in the Dirichlet case. We

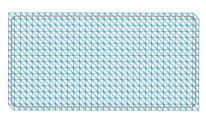

(a) Dirichlet boundary conditions are imposed on the left edge of the rectangle. Γ_N is marked in red, and a force g is acting on it (indicated by the green arrow).

(b) The overlapping triangulation $\mathcal{T}$ (cyan) together with the boundary $\partial\mathcal{O}$ (red).

Fig. 2.10: (a) shows a sketch of the set-up for a small example problem in two dimensions. (b) shows the triangulation $\mathcal{T}$ and the boundary $\partial\mathcal{O}$ for this setting.

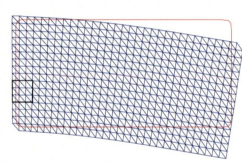
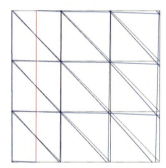

(a) The deformation u obtained as the solution to the elasticity PDE on the overlapping triangulation $\mathcal{T}$ is drawn in blue.

(b) Magnification of the marked area in (a). The dashed lines indicate the undeformed grid.

Fig. 2.11: The solution u to the elasticity PDE (1.7) for the set-up depicted in Fig. 2.10(a) is shown in (a). A small portion of the grid intersected by the Dirichlet boundary (marked by a black square) is magnified in (b). It can be seen that the Dirichlet boundary conditions are indeed approximately satisfied on the interface.

evaluated the relative error for one of our test instances in Chapter 5 in the stress when refining the underlying grid once, i.e. we computed the number

$$\frac{\int_{\mathcal{O}}\left[Ae\left(U_h(\mathcal{O})\right) - Ae\left(U_{\frac{h}{2}}(\mathcal{O})\right)\right]^2 \mathrm{d}x}{\int_{\mathcal{O}}\left[Ae\left(U_h(\mathcal{O})\right)\right]^2 \mathrm{d}x}, \quad (2.24)$$

2.1 Composite Finite Elements

with $\mathscr{O}$ being the stochastically optimal, discrete shape from Fig. 5.2 on page 103 subjected to $g(\omega_1)$ as given in the same figure. We chose $h = 2^{-8}$ which corresponds to a uniform grid with $(2^8 + 1) \times (2^8 + 1)$ nodes. The solution obtained on the grid is denoted by $U_h(\mathscr{O})$. Refining that grid once yields a grid with $(2^9 + 1) \times (2^9 + 1)$ nodes, and the corresponding solution $U_{\frac{h}{2}}(\mathscr{O})$. Precisely, we obtained a relative error of about 0.25% for the number defined in (2.24).

2.1.4 Computation of the System Matrix and the Right-Hand Side Vector

In this section, we briefly describe how we computed the integrals appearing in the representation of the local element matrices and vectors (2.10). Because of (2.11), it suffices to consider integrals over subtriangles to be able to compute $B_{i,j}^{l,k}(\tau)$ and $f_i^l(\tau)$. The computation of $g_i^l(\tau)$ only involves triangles that are intersected by the Neumann boundary Γ_N. Therefore, we look at a fixed triangle t which is in $\mathscr{G}(\tau)$ for some $\tau \in \mathscr{T}$, and at a fixed triangle τ_N that is intersected by Γ_N, respectively. We assume that f and g are also given in V_h^{CFE}, which allows the following representations according to (2.9):

$$f = \sum_{i=1}^{N} \sum_{j=1}^{2} F_i^j b_i^{\mathrm{CFE},j} \quad \text{and} \quad g = \sum_{i=1}^{N} \sum_{j=1}^{2} G_i^j b_i^{\mathrm{CFE},j}.$$

Then, one can easily check that the integrals occurring in the expressions for $B_{i,j}^{l,k}(t)$, $f_i^l(t)$, and $g_i^l(\tau_N)$ can be broken down to several integrals of the following form (neglecting any coefficients):

$$\int_t b_{j,l}(x,y) b_{i,k}(x,y) \, d(x,y), \quad 1 \leq k,l \leq 2, \, 1 \leq i,j \leq 3,$$

$$\int_t b_j(x,y) b_i(x,y) \, d(x,y), \quad 1 \leq i,j \leq 3,$$

$$\int_{\tau_N \cap \Gamma_N} b_j(x,y) b_j(x,y) \, ds(x,y), \quad 1 \leq i,j \leq 3.$$

Only the scalar basis functions and their partial derivatives, respectively, are involved in these integrals. The standard procedure (cf. [Bra03]) to compute the above integrals then is to transform the triangle t to the so-called *reference triangle* τ_{ref}, which has the vertices $(0,0),(1,0),(0,1)$, and to make use of the *change of variables formula* for integrals. Finally, the resulting integrals over τ_{ref} are computed using simplicial quadrature rules (cf. for example [DR75, Str71]).

3 Stochastic Programming Perspective

The objective of this chapter is to formulate a *stochastic* shape optimization problem, which allows for uncertain volume forces and surface loads. This reflects applications where the actual acting forces are not known in advance, but a decision maker has to make a design decision in advance nonetheless. A few of these applications have been mentioned already in Chapter 1. We will then show that this problem is very similar to a two-stage stochastic linear program (cf. Section 1.3), where first-stage decisions also have to be taken in a non-anticipative way.

It will turn out that the variational formulation of the elasticity PDE stated in Theorem 1.5 on page 11 can be considered a natural counterpart of the second-stage minimization problem (1.25), at least formally. Of course, as the minimization problem in Theorem 1.5 admits a unique solution, this point of view might seem rather artificial. After all, there is no actual decision making necessary — in fact, not even possible — in the second-stage problem unlike in the linear second-stage problem. However, this very formulation emphasizes intriguing similarities between the two types of problems and allows us to transfer some ideas from the finite dimensional case to our infinite dimensional random shape optimization problem. In particular, a notion of duality will play an important role which is worked out in Section 3.1. Additionally, this perspective enables us to formulate stochastic shape optimization problems involving risk measures (cf. Section 1.3.2 on page 22) coming from finite dimensional stochastic programming, in particular the *expected excess* (see Section 3.3) and *excess probability* (see Section 3.4).

An important algorithmic shortcut is achieved by assuming a special structure for the random forces without loss of generality. More precisely, we assume that there are finitely many basis volume forces and surface loads. These basis forces are deterministic. Actual scenarios are then formed by combining these basis forces linearly, where the coefficients in the linear combinations are uncertain, i.e. random variables. This leads to a reformulation of our stochastic shape optimization problem which immediately reveals an efficient way to solve it. Thanks to linearity, we only need to solve as many elasticity PDEs as there are basis forces; a number which is obviously totally independent of the number of scenarios. Details about this can be found in Section 3.2.

Note that there are approaches in shape optimization that generalize the single load case by taking multiple loads into account (cf. e.g. [AJ05, BTKNZ99] and references therein). In these so-called *multiload* approaches, a fixed, usually small, number of different loading configurations, which act on the elastic body simultaneously, are considered. They require the solution of PDEs for each loading configuration which constitutes the essential difference compared to the approach we present in this chapter, as we only need to solve PDEs for the basis forces. The number of possible forces can be rather large in our case, which makes it possible to approximate a continuous distribution of forces.

3.1 Stochastic Shape Optimization Problem

We now develop a formulation for a shape optimization problem which may include uncertain parameters. As mentioned before, these uncertain parameters are the volume forces and surface loads specifically. Therefore, we indicate the randomness of f and g by writing $f(\omega)$ and $g(\omega)$, respectively. As explained in Section 1.3, the latter are random variables depending on the outcome of random experiments $\omega \in \Omega = \{\omega_1, \ldots, \omega_S\}$. In that sense, we assume that whenever there is a ω, it is understood to belong to the set of scenarios $\{\omega_1, \ldots, \omega_S\}$, even without explicitly stating it. The first thing to do now is to include the random forces into the formulation of the elasticity PDE (1.7). With this in mind, one only needs to modify the right-hand side at the appropriate spots leading to

$$\begin{cases} -\operatorname{div}(Ae(u)) = f(\omega) & \text{in } \mathcal{O}, \\ u = 0 & \text{on } \Gamma_D, \\ (Ae(u))n = g(\omega) & \text{on } \Gamma_N, \\ (Ae(u))n = 0 & \text{on } \Gamma_0. \end{cases} \quad (3.1)$$

Obviously, the solution u to (3.1) does not only depend on $\mathcal{O}$ now, but also on ω. We indicate this by writing $u = u(\mathcal{O}; \omega)$.

Next, we need to adapt some definitions introduced in Section 1.1.1 to exhibit the randomness, i.e. the dependence on ω.

Notation 3.1. *In complete analogy to Definition 1.2 on page 6 we introduce the following notations in the context of our random shape optimization problem for $u, \psi, \vartheta \in V$:*

- $l(\mathcal{O}, \vartheta; \omega) := \int_{\mathcal{O}} f(\omega) \cdot \vartheta \, dx + \int_{\Gamma_N} g(\omega) \cdot \vartheta \, ds$ *in correspondence to* (1.10), *and*

- $E(\mathcal{O}, u; \omega) := \frac{1}{2} A(\mathcal{O}, u, u) - l(\mathcal{O}, u; \omega)$ *corresponding to* (1.9).

3.1 Stochastic Shape Optimization Problem

With these notations, it is immediately evident that results concerning the existence of a unique solution to (3.1) for a fixed $\omega \in \Omega$ again follow from Theorem 1.5. Likewise, the statements in Lemma 1.3, Theorem 1.4, and Lemma 1.6 hold analogously. Furthermore, we can now easily introduce our stochastic shape optimization model we are aiming to solve with Definition 1.10 in mind.

Definition 3.2 (Stochastic shape optimization model). *We consider the stochastic shape optimization problem given by*

$$\min\{\mathbf{J}(\mathcal{O};\omega) : \mathcal{O} \in \mathcal{U}_{ad}, \quad dE(\mathcal{O}, u(\mathcal{O};\omega); \omega; \psi) = 0, \forall \psi \in V\}, \tag{3.2}$$

with

$$\begin{aligned}\mathbf{J}(\mathcal{O};\omega) = J(\mathcal{O}, u(\mathcal{O};\omega)) &= \int_{\mathcal{O}} j(x, u(\mathcal{O};\omega))\,dx + \int_{\partial\mathcal{O}} k(x, u(\mathcal{O};\omega))\,ds \\ &\quad + \alpha \int_{\mathcal{O}} 1\,dx + \beta \int_{\partial\mathcal{O}} 1\,ds,\end{aligned} \tag{3.3}$$

and, as before, $\mathcal{U}_{ad} = \{\mathcal{O} \subseteq D : \mathcal{O}$ has Lipschitz boundary and $\partial\mathcal{O}$ satisfies (1.6)$\}$.

Remark 3.3. *We primarily still have the compliance* (1.14) *and the least square error functional* (1.15) *in mind as special cases for the general objective function given by* (3.3), *particularly its integrands $j(x,u)$ and $k(x,u)$. The only difference is that u is now solution to the elasticity PDE with stochastic forces, namely* (3.1). *In principle, different choices are possible and conceivable. However, we impose one restriction on j and k which turns out to be crucial for our approach: We assume from now on that j and k depend linearly or quadratically on u such that $j(.,u), j_{,u}(.,u) \in L^2(\mathcal{O})$, and $k(.,u), k_{,u}(.,u) \in L^2(\partial\mathcal{O})$. The reason for this is that the derivatives with respect to u of j and k are then constant or linear. Obviously, the compliance* (1.14) *and the quadratic objective functional* (1.15) *satisfy that requirement.*

3.1.1 Two-Stage Stochastic Shape Optimization Problem

We demonstrate in this section how the stochastic shape optimization problem (3.2) can be apprehended as a two-stage stochastic optimization problem in the sense described in Section 1.3. To see this, we first rewrite (3.2) by means of Theorem 1.5 and Lemma 1.6 as

$$\min\left\{J(\mathcal{O}, u(\mathcal{O};\omega)) : u(\mathcal{O};\omega) = \arg\min_{v \in V} E(\mathcal{O}, v; \omega), \mathcal{O} \in \mathcal{U}_{ad}\right\}. \tag{3.4}$$

Next, we replace the expression $J(\mathcal{O}, u(\mathcal{O}; \omega))$ in (3.4) by its definition according to (3.3), deliberately changing the order of the terms occurring in (3.3), to obtain the following problem formulation:

$$\min \left\{ \alpha \int_{\mathcal{O}} 1 \, dx + \beta \int_{\partial \mathcal{O}} 1 \, ds + \int_{\mathcal{O}} j(x, u(\mathcal{O}; \omega)) \, dx + \int_{\partial \mathcal{O}} k(x, u(\mathcal{O}; \omega)) \, ds : \right.$$
$$\left. u(\mathcal{O}; \omega) = \arg\min_{v \in V} E(\mathcal{O}, v; \omega), \ \mathcal{O} \in \mathcal{U}_{ad} \right\}. \quad (3.5)$$

At this point — being completely aware of the clash in notations — it makes sense to take a closer look at the general linear two-stage stochastic program introduced in Section 1.3, in particular the formulation given by (1.21):

$$\min_x \left\{ c^T x + \min_y \{ q^T y : Wy = z(\omega) - Tx, y \in Y \} : x \in X \right\}.$$

Instead of explicitly putting the inner minimization problem, i.e. the second-stage problem, in the objective function, we can also write the above problem in a slightly different way:

$$\min_x \left\{ c^T x + q^T \bar{y} : \bar{y} \in \arg\min_y \{ q^T y : Wy = z(\omega) - Tx, y \in Y \}, x \in X \right\}. \quad (3.6)$$

Now we can directly compare our stochastic shape optimization problem (3.5) with the general linear two-stage stochastic model (3.6). We find that (3.5) can be seen as a two-stage optimization problem with the following counterparts:

$$x \hat{=} \mathcal{O},$$
$$y \hat{=} u,$$
$$c^T x \hat{=} \alpha \int_{\mathcal{O}} 1 \, dx + \beta \int_{\partial \mathcal{O}} 1 \, ds,$$
$$q^T y \hat{=} \int_{\mathcal{O}} j(x, u(\mathcal{O}; \omega)) \, dx + \int_{\partial \mathcal{O}} k(x, u(\mathcal{O}; \omega)) \, ds.$$

So the first-stage decision in the case of the random shape optimization problem is the decision on the shape $\mathcal{O}$. This decision has to be taken without knowing the actual forces, in other words non-anticipatively. Only those costs that arise as a consequence of the choice of the shape alone constitute the first-stage objective functional. In our case it is clear from (3.3) that $\alpha \int_{\mathcal{O}} 1 \, dx + \beta \int_{\partial \mathcal{O}} 1 \, ds$ plays that role here. This is also the reason why we changed the order of the terms in the objective function in (3.5), as then the similarities between (3.5) and (3.6) become even more evident.

3.1 Stochastic Shape Optimization Problem

The remaining part of the objective (3.3) forms the analogue to the second-stage objective function since it involves the deformation u — and u itself can be seen as the second-stage variable. Just like $\bar{y}$ in the linear case (3.6), u is a solution of a specific minimization problem, namely the one coming from the variational formulation of the elasticity PDE (see Thm. 1.5). The only difference, however, is that the solution of this *inner* minimization problem is unique in (3.5), whereas in the linear case (3.6) one can typically select a best recourse action among many feasible options. We have noted this issue already in the beginning of this chapter.

Summarizing, we can say that (3.5) can be considered a two-stage stochastic optimization problem in the sense that one has to decide on the shape $\mathcal{O}$ before knowing the realizations of the random forces $f(\omega)$ and $g(\omega)$. Afterwards, once the actual forces can be observed, one can compute the resulting deformation, given $\mathcal{O}$ and these forces. This leads to the same type of information constraint we have seen already in the linear case in (1.19):

$$\text{decide } \mathcal{O} \longrightarrow \text{observe } f(\omega), g(\omega) \longrightarrow \text{compute } u = u(\mathcal{O}; \omega).$$

According to (3.5), we have to find a "best" member in the family of random variables $\{\bar{G}(\mathcal{O}; \omega) : \mathcal{O} \in \mathcal{U}_{ad}\}$ with

$$\bar{G}(\mathcal{O}; \omega) := \alpha \int_{\mathcal{O}} 1\,dx + \beta \int_{\partial \mathcal{O}} 1\,ds + \int_{\mathcal{O}} j(x, u(\mathcal{O}; \omega))\,dx + \int_{\partial \mathcal{O}} k(x, u(\mathcal{O}; \omega))\,ds,$$

such that $u(\mathcal{O}; \omega)$ solves (3.1). Recall Section 1.3 where we discussed that these random variables need to be ranked by some criterion in order to make sense and to become computationally sound. In our first model, we simply rank them by their expectations (see Section 3.2). Later, we also discuss models involving the expected excess (cf. Section 3.3) and the excess probability (cf. Section 3.4) as risk measures.

Definition 3.4 (Expectation-based model). *Analogously to (1.22), the random expectation-based shape optimization problem reads as*

$$\min\{\mathbb{E}_\omega\left(\bar{G}(\mathcal{O}; \omega)\right) : \mathcal{O} \in \mathcal{U}_{ad}\}. \tag{3.7}$$

With $\bar{G}$ defined as stated above, the objective can be written as

$$\alpha \int_{\mathcal{O}} 1\,dx + \beta \int_{\partial \mathcal{O}} 1\,ds + \mathbb{E}_\omega\left(\int_{\mathcal{O}} j(x, u(\mathcal{O}; \omega))\,dx + \int_{\partial \mathcal{O}} k(x, u(\mathcal{O}; \omega))\,ds\right), \tag{3.8}$$

where $u(\mathcal{O}; \omega)$ satisfies the constraint $u(\mathcal{O}; \omega) = \arg\min_{v \in V} E(\mathcal{O}, v; \omega)$.

3.1.2 Dual Problem and Saddle Point Formulation

In the linear finite dimensional case, it turns out to be useful to consider the linear programming dual problem of the second-stage minimization problem. This is indicated in Section 1.3.1, see in particular (1.26). Of course, the linear programming duality theorem (cf. Theorem A.11 on page 125) does not apply to our random shape optimization problem (3.7) since it is neither finite dimensional nor linear. However, in nonlinear optimization there is also a notion of duality, namely *Lagrangian duality* (see for example [Rus06, Chapter 4]). In [ET76, Chapter VI] they demonstrate how to construct dual problems also for (convex) variational problems, which also involves Lagrangian functions. We have mentioned already in Section 1.2 that Lagrangian functions also play an important role in optimal control theory (cf. [Trö05]), and that shape optimization problems can be considered as such. This leads to a saddle point formulation (cf.[DJPZ01, p. 421]) and the adjoint state. With the following remark we would like to motivate this approach in terms of proceeding analogously to the linear stochastic programming formulation, in particular the linear programming dual formulation of the second-stage problem.

Remark 3.5. *Let us consider a linear optimization problem given in the form of the second-stage problem* (1.25), *i.e.*

$$\min\{q^T y : Wy = v, y \geq 0\}. \tag{3.9}$$

Then we can apply the same techniques used for nonlinear optimization problems to obtain a Lagrangian dual problem (cf. [Rus06, p. 160 ff., in particular Example 4.2]) to (3.9). *In doing so, we get*

$$L(y,z) := q^T y + z^T (Wy - v), \; \forall y \geq 0, \forall z \in \mathbb{R}^l$$

as the Lagrangian. We denote the dual function by F, which is given for all $z \in \mathbb{R}^l$ by

$$F(z) := \inf\{L(y,z) : y \geq 0\}.$$

The (Lagrangian) dual problem is then

$$\sup\left\{F(z) : z \in \mathbb{R}^l\right\}. \tag{3.10}$$

We observe that the Lagrangian can also be written as

$$L(y,z) = q^T y + z^T (Wy - v)$$

3.1 Stochastic Shape Optimization Problem

$$= \left(q^T + z^T W\right) y - z^T v.$$

Therefore, the dual function F can be expressed as

$$F(z) = -z^T v + \inf\left\{\left(q^T + z^T W\right) y : y \geq 0\right\}.$$

We can deduce from this formulation that F can only take finite values, i.e. $F(z) > -\infty$, *if* $\left(q^T + z^T W\right) \geq 0$ *holds. This can be seen as follows: Suppose, there is a component* $i \in \{1, \ldots, m\}$ *with* $q_i + (z^T W)_i < 0$. *Then we could define feasible points* $y(t) := (t\delta_{1i}, \ldots, t\delta_{mi}) \in \mathbb{R}^m$ *for all* $t \in \mathbb{R}, t \geq 0$. *Letting* $t \geq 0$ *tend to* $+\infty$ *would then yield* $\left(q^T + z^T W\right) y(t) \longrightarrow -\infty$, *and consequently* $F(z) = -\infty$.

$F(z) = -\infty$ *is clearly not desirable for the dual problem* (3.10), *as we want to maximize* $F(z)$ *in that problem. Hence an optimal* $z \in \mathbb{R}^l$ *has to satisfy* $q^T + z^T W \geq 0$, *or equivalently* $W^T(-z) \leq q$. *The dual function then reduces to* $F(z) = -z^T v + 0$ *for all* $z \in \mathbb{R}^l$, *and the dual problem* (3.10) *can be formulated as*

$$\max\left\{-z^T v : W^T(-z) \leq q\right\},$$

which coincides with the linear programming dual problem in (1.25) *after replacing*[1] $-z$ *by u.*

Before we describe the construction of a Lagrangian functional for our problem, let us briefly note that we can also write problem (3.5) slightly differently to match formulation (1.21); that way there actually are *inner* and *outer* minimization problems. This reformulation is trivial, and valid because u is the unique minimizer of $E(\mathcal{O}, v; \omega)$ (cf. constraints in (3.5)). With (1.21) in mind, we rewrite (3.5) as

$$\min\left\{\alpha \int_{\mathcal{O}} 1\,\mathrm{d}x + \beta \int_{\partial\mathcal{O}} 1\,\mathrm{d}s + \bar{\Phi}(\mathcal{O}; f(\omega), g(\omega)) : \mathcal{O} \in \mathcal{U}_{\text{ad}}\right\},$$

with

$$\bar{\Phi}(\mathcal{O}; f(\omega), g(\omega)) := \min\left\{\int_{\mathcal{O}} j(x, u(\mathcal{O}; \omega))\,\mathrm{d}x + \int_{\partial\mathcal{O}} k(x, u(\mathcal{O}; \omega))\,\mathrm{d}s : \right.$$
$$\left. \mathrm{d}E(\mathcal{O}, u(\mathcal{O}; \omega); \omega; \psi) = 0, \, \forall \psi \in V\right\}. \quad (3.11)$$

$f(\omega)$ and $g(\omega)$ come into play in the above formulation through $E(\mathcal{O}, v; \omega)$ via its definition in Notation 3.1. Note that we included the condition that u solves the (random) elasticity PDE (3.1) again as an equality constraint as in (3.2).

---
[1] z has no sign restriction, so it is equivalent to write $u \in \mathbb{R}^l$ instead of $-z \in \mathbb{R}^l$.

For convenience, we denote the second-stage objective functional by $\bar{\mathbf{J}}(\mathscr{O};\omega) = \bar{J}(\mathscr{O},u(\mathscr{O};\omega))$, i.e.

$$\bar{J}(\mathscr{O},u(\mathscr{O};\omega)) = \int_{\mathscr{O}} j(x,u(\mathscr{O};\omega))\,\mathrm{d}x + \int_{\partial\mathscr{O}} k(x,u(\mathscr{O};\omega))\,\mathrm{d}s.$$

From the proof of Lemma 1.6 we know together with Notation 3.1 that

$$\mathrm{d}E(\mathscr{O},u(\mathscr{O};\omega);\omega;\psi) = A(\mathscr{O},u,\psi) - l(\mathscr{O},\psi;\omega) \qquad (3.12)$$

$$= \int_{\mathscr{O}} Ae(u):e(\psi)\,\mathrm{d}x - \int_{\mathscr{O}} f(\omega)\cdot\psi\,\mathrm{d}x - \int_{\Gamma_N} g(\omega)\cdot\psi\,\mathrm{d}s.$$

Then we can introduce a Lagrangian functional as demonstrated in [DJPZ01, p. 422 ff.] and [AJT04] by introducing a *Lagrange multiplier* function which is called *adjoint state* ψ:

$$L(\mathscr{O},\varphi,\psi;\omega) := \bar{J}(\mathscr{O},\varphi) + \mathrm{d}E(\mathscr{O},\varphi;\omega;\psi),\ \forall\varphi,\psi\in V. \qquad (3.13)$$

For the sake of readability, we also shortly write $j(u)$ and $k(u)$ instead of $j(x,u)$ and $k(x,u)$, respectively, for the integrands appearing in the objective (3.3). Then we obtain for all $\varphi,\psi\in V$, taking (3.12) and (3.13) into account

$$L(\mathscr{O},\varphi,\psi;\omega) = \int_{\mathscr{O}} j(\varphi)\,\mathrm{d}x + \int_{\partial\mathscr{O}} k(\varphi)\,\mathrm{d}s + \int_{\mathscr{O}} Ae(\varphi):e(\psi)\,\mathrm{d}x$$

$$- \int_{\mathscr{O}} f(\omega)\cdot\psi\,\mathrm{d}x - \int_{\Gamma_N} g(\omega)\cdot\psi\,\mathrm{d}s. \qquad (3.14)$$

Now there are two possible cases for the choice of $\varphi\in V$ which are worth taking a closer look at. The first one is the choice $\varphi = u(\mathscr{O};\omega)$, which in a sense could be considered the "right" choice. By Lemma 1.6, this choice yields

$$L(\mathscr{O},u(\mathscr{O};\omega),\psi;\omega) = \bar{J}(\mathscr{O},u(\mathscr{O};\omega)).$$

Note that in this case the value of $L(\mathscr{O},\varphi,\psi;\omega)$ is totally independent of ψ.

On the contrary, if $\varphi\in V$ is chosen such that $\varphi\neq u(\mathscr{O};\omega)$, we can make the following observations: The condition $\mathrm{d}E(\mathscr{O},u;\omega;\psi) = 0$, $\forall\psi\in V$, is equivalent to (1.11) according to the proof of Lemma 1.6. Furthermore, (1.11) admits exactly one solution by Theorem 1.5. Obviously, as $\varphi\neq u(\mathscr{O};\omega)$, our φ is not the solution to (1.11), and must therefore give a value $\mathrm{d}E(\mathscr{O},\varphi;\omega;\psi)\neq 0$ for at least one $\psi\in V$. From (3.12) we know that $\mathrm{d}E(\mathscr{O},\varphi;\psi;\omega)$ is linear in ψ. Thus, since V is a linear space, $L(\mathscr{O},\varphi,\psi;\omega)$ may become arbitrarily big in this second case.

3.1 Stochastic Shape Optimization Problem

To sum up, we have just shown that

$$\sup_{\psi \in V} L(\mathcal{O}, \varphi, \psi; \omega) = \begin{cases} \bar{J}(\mathcal{O}, u(\mathcal{O}; \omega)) & \text{if } \varphi = u(\mathcal{O}; \omega), \\ +\infty & \text{if } \varphi \neq u(\mathcal{O}; \omega). \end{cases} \quad (3.15)$$

Consequently, the objective functional $\bar{J}(\mathcal{O}, u(\mathcal{O}; \omega))$ can be expressed as

$$\bar{\mathbf{J}}(\mathcal{O}; \omega) = \min_{\varphi \in V} \sup_{\psi \in V} L(\mathcal{O}, \varphi, \psi; \omega). \quad (3.16)$$

Following the constructional ideas demonstrated in Remark 3.5, we obtain the dual problem

$$\max \{ \bar{F}(\psi; \omega) : \psi \in V \},$$

with $\bar{F}(\psi; \omega) := \inf \{ L(\mathcal{O}, \varphi, \psi; \omega) : \varphi \in V \}$.

(3.16) shows that the evaluation of the objective function $\bar{\mathbf{J}}(\mathcal{O}; \omega)$ for a given shape $\mathcal{O}$ is closely related to finding a saddle point for the Lagrangian L, if it satisfies appropriate conditions. If the Lagrangian L satisfies certain assumptions, a saddle point of L is characterized by the stationarity of L (cf. Ekeland and Temam [ET76, Chapter VI, in particular Proposition 1.6]). In our case, L is given by (3.14), and the necessary requirements arise from simple observations which we summarize in the following lemma.

Lemma 3.6. *Let the Lagrangian $L(\mathcal{O}, \varphi, \psi; \omega)$ be given as in (3.14). Then the following holds:*

(i) *L is continuous with respect to the variable φ. If $j(.)$ and $k(.)$ are convex, L is also convex with respect to the variable φ.*

(ii) *L is concave and continuous with respect to the variable ψ.*

(iii) *The space V is convex and closed.*

(iv) *For all $\varphi \in V$, $\psi \mapsto L(\mathcal{O}, \varphi, \psi; \omega)$ is Gâteaux-differentiable.*

(v) *For all $\psi \in V$, $\varphi \mapsto L(\mathcal{O}, \varphi, \psi; \omega)$ is Gâteaux-differentiable.*

Proof.

(i) If j and k are convex, all terms occurring in (3.14) are convex with respect to φ, and hence L must be convex in φ. The continuity follows from the assumptions on j and k in Remark 3.3, the Cauchy-Schwarz inequality (Theorem A.10), and Lemma 1.3.

(ii) Similar to (i), L is an affine function in ψ.

(iii) This is a direct consequence from the definition of the space V (cf. Definition 1.2 and Notation A.5).

The Gâteaux-differentiability[2] of L with respect to φ and ψ follows from the special assumptions on $j(\varphi)$ and $k(\varphi)$ and Lebesgue's dominated convergence theorem (cf. [DS57, Els02]). □

The next lemma provides some useful conditions to express the objective $\bar{\mathbf{J}}(\mathcal{O};\omega)$ by means of the Lagrangian L. In case $j(\varphi)$ and $k(\varphi)$ are convex, these amount to saddle point conditions.

Lemma 3.7. $(u,p) \in V \times V$ satisfies

$$\bar{\mathbf{J}}(\mathcal{O};\omega) = L(\mathcal{O},u,p;\omega),$$

if

$$\left\langle \frac{\partial L}{\partial \varphi}(\mathcal{O},u,p;\omega),\Theta \right\rangle = 0, \forall \Theta \in V, \tag{3.17}$$

$$\left\langle \frac{\partial L}{\partial \psi}(\mathcal{O},u,p;\omega),\Theta \right\rangle = 0, \forall \Theta \in V. \tag{3.18}$$

Proof. If $j(\varphi)$ and $k(\varphi)$ are convex, this follows immediately from Lemma 3.6 and [ET76, Proposition 1.6, p. 169]. In this case, $(u,p) \in V \times V$ is a saddle point of L, and the statement of the lemma also holds conversely.

If (3.17) and (3.18) hold, u and p are uniquely determined as the solutions to (3.1) and (3.20), respectively, anticipating what follows in the rest of this section[3]. Then we know from (3.15) that

$$\sup_{\psi \in V} L(\mathcal{O},u,\psi;\omega) = \bar{\mathbf{J}}(\mathcal{O};\omega).$$

From Lemma 3.6 we know that L is concave with respect to ψ. Therefore, (3.18) is the sufficient optimality condition satisfied by p, which means that $L(\mathcal{O},u,p;\omega) = \sup_{\psi \in V} L(\mathcal{O},u,\psi;\omega)$, which completes the proof. □

---
[2] See e.g. [ET76, p. 23] for the definition.
[3] We will show in the rest of this section that (3.17) corresponds to the weak formulation of an elasticity PDE, namely the so-called adjoint problem (cf. Definition 3.8). Likewise, (3.18) corresponds to the original elasticity PDE (3.1), such that u and p are uniquely determined.

3.1 Stochastic Shape Optimization Problem

Let us compute the above conditions, starting with (3.17):

$$0 = \left\langle \frac{\partial L}{\partial \varphi}(\mathscr{O}, u, p; \omega), \Theta \right\rangle = \int_{\mathscr{O}} j'(u) \cdot \Theta \, dx + \int_{\partial \mathscr{O}} k'(u) \cdot \Theta \, ds$$
$$+ \int_{\mathscr{O}} Ae(\Theta) : e(p) \, dx. \qquad (3.19)$$

Integrating by parts yields further (recall the proof of Theorem 1.4):

$$0 = \int_{\mathscr{O}} (-\operatorname{div}(Ae(p))) \cdot \Theta \, dx + \int_{\partial \mathscr{O}} ((Ae(p))n) \cdot \Theta \, ds$$
$$+ \int_{\mathscr{O}} j'(u) \cdot \Theta \, dx + \int_{\partial \mathscr{O}} k'(u) \cdot \Theta \, ds$$
$$= \int_{\mathscr{O}} (-\operatorname{div}(Ae(p)) + j'(u)) \cdot \Theta \, dx + \int_{\Gamma_N \cup \Gamma_0} ((Ae(p))n + k'(u)) \cdot \Theta \, ds.$$

This last expression makes it clear that (3.17) is a PDE in weak form. In the following, we derive its strong formulation: At first we take Θ with compact support in $\mathscr{O}$ to obtain

$$-\operatorname{div}(Ae(p)) = -j'(u) \quad \text{in } \mathscr{O}.$$

Then we vary the trace of Θ on $\Gamma_N \cup \Gamma_0$ which leads to the Neumann boundary conditions

$$(Ae(p))n = -k'(u) \quad \text{on } \Gamma_N \cup \Gamma_0.$$

Finally, since we are looking for p in V, we have

$$p = 0 \quad \text{on } \Gamma_D.$$

The resulting PDE is called *adjoint state equation*. For future references, we summarize it in the following definition.

Definition 3.8. *The* adjoint problem *reads as*

$$\begin{cases} -\operatorname{div}(Ae(p)) = -j'(u) & \text{in } \mathscr{O} \\ p = 0 & \text{on } \Gamma_D \\ (Ae(p))n = -k'(u) & \text{on } \Gamma_N \cup \Gamma_0. \end{cases} \qquad (3.20)$$

Its solution p is called the adjoint state.

Remark 3.9. *When comparing (3.20) to the elasticity PDE (1.7), one realizes that (3.20) is also an elasticity PDE and therefore admits an unique solution p.*

Similarly, we can compute (3.18):

$$0 = \left\langle \frac{\partial L}{\partial \psi}(\mathscr{O}, u, p; \omega), \Theta \right\rangle = \int_{\mathscr{O}} Ae(u) : e(\Theta)\, dx - \int_{\mathscr{O}} f(\omega) \cdot \Theta\, dx$$
$$- \int_{\Gamma_N} g(\omega) \cdot \Theta\, ds. \tag{3.21}$$

Again, we integrate by parts and obtain

$$0 = \int_{\mathscr{O}} (-\operatorname{div}(Ae(u)) - f(\omega)) \cdot \Theta\, dx + \int_{\Gamma_0} (\sigma n) \cdot \Theta\, ds$$
$$+ \int_{\Gamma_N} (\sigma n - g(\omega)) \cdot \Theta\, ds$$

by the proof of Theorem 1.4. This last expression looks almost exactly like the one at the end of the proof of Theorem 1.4 — except for the now random forces $f(\omega)$ and $g(\omega)$. This shows that the strong PDE corresponding to (3.18) is the random, original elasticity PDE (3.1).

In other words, we have shown that the unique solution $(u, p) \in V \times V$ of the system consisting of (3.1) and (3.20) coincides with the therefore unique saddle point of $L(\mathscr{O}, \varphi, \psi; \omega)$ in $V \times V$, if $j(\varphi)$ and $k(\varphi)$ are convex[4]. Otherwise, the objective functional $\bar{\mathbf{J}}(\mathscr{O}; \omega)$ can still be expressed by means of the Lagrangian L according to Lemma 3.7. The adjoint state p will play an important role later in Chapter 4 as it appears in the *shape derivative* (see. Section 4.2).

Remark 3.10. *The adjoint PDE* (3.20) *simplifies significantly for the special case of the compliance objective* (1.14):

$$\mathbf{J}_1(\mathscr{O}; \omega) = \int_{\mathscr{O}} f(\omega) \cdot u\, dx + \int_{\Gamma_N} g(\omega) \cdot u\, ds.$$

In particular, we then have

$$j(u) = f(\omega) \cdot u, \quad k(u) = (g(\omega) \cdot u)\chi_{\Gamma_N},$$

where χ_M denotes the characteristic function of a set M, i.e. $\chi_M(x) = 1$ if $x \in M$ and $\chi_M(x) = 0$ otherwise. Hence

$$j'(u) = f(\omega), \quad k'(u) = \begin{cases} 0 & \text{on } \Gamma_0, \\ g(\omega) & \text{on } \Gamma_N. \end{cases}$$

---

[4] This is the case for the objective functionals we have in mind, namely (1.14) and (1.15).

This means that (3.20) for the compliance objective reads as

$$\begin{cases} -\operatorname{div}(Ae(p)) = -f(\omega) & \text{in } \mathcal{O} \\ p = 0 & \text{on } \Gamma_D \\ (Ae(p))n = 0 & \text{on } \Gamma_0 \\ (Ae(p))n = -g(\omega) & \text{on } \Gamma_N. \end{cases}$$

Comparing this to (3.1) shows that $p = -u$ in this special case. In that sense, the problem is self-adjoint, and the adjoint state need not be computed explicitly which saves time in the numerical algorithm since it would require another solution of an elasticity PDE.

3.2 Reformulation and Solution Plan for the Expectation-Based Model

This section is particularly dedicated to the expectation-based stochastic shape optimization model (3.7) from Definition 3.4 on page 53. The special structure of the random forces and the consequent algorithmic shortcut that are presented here, however, also apply for the stochastic models to be discussed in the next two sections 3.3 and 3.4 involving risk measures.

We mentioned this special structure of the random forces $f(\omega)$ and $g(\omega)$ already vaguely in the beginning of this chapter, and we will now give the full particulars in the subsequent definition.

Definition 3.11 (Structure of random forces). *Let $f_1, \ldots, f_{K_1} \in L^2(D; \mathbb{R}^2)$ be finitely many deterministic volume forces, and let $g_1, \ldots, g_{K_2} \in H^1(D; \mathbb{R}^2)$ be finitely many deterministic surface loads. In the sequel, we will also refer to these volume forces and surface loads as* basis forces, *as we assume that the forces in the actual scenarios are linear combinations of these basis forces where the coefficients vary from scenario to scenario, i.e. the coefficients are considered random. More precisely, this means that $f(\omega)$ and $g(\omega)$ allow the following representations:*

$$f(\omega) = \sum_{i=1}^{K_1} c_i^f(\omega) f_i, \qquad g(\omega) = \sum_{j=1}^{K_2} c_j^g(\omega) g_j, \qquad (3.22)$$

with the uncertain coefficients $c_i^f(\omega) \in \mathbb{R}$, $i = 1, \ldots, K_1$, and $c_j^g(\omega) \in \mathbb{R}$, $j = 1, \ldots, K_2$, respectively.

Remark 3.12. *In the special case that the coefficients $c_i^f(\omega)$ and $c_j^g(\omega)$ from the previous Definition 3.11 are all greater than 0 and add up to 1, i.e.*

$$\sum_{i=1}^{K_1} c_i^f(\omega) = 1, \qquad \sum_{j=1}^{K_2} c_j^g(\omega) = 1,$$

these coefficients can be interpreted as probabilities themselves[5]. A scenario ω then differs from another scenario ω' by the probability estimates for the forces f_i and g_j. In general, of course, this situation need not be the case.

Suppose we have u and p that satisfy

$$0 = \int_{\mathcal{O}} j'(u) \cdot \Theta \, dx + \int_{\partial \mathcal{O}} k'(u) \cdot \Theta \, ds + \int_{\mathcal{O}} Ae(\Theta) : e(p) \, dx, \quad \forall \Theta \in V, \quad (3.23a)$$

$$0 = \int_{\mathcal{O}} Ae(u) : e(\Theta) \, dx - \int_{\mathcal{O}} f_i \cdot \Theta \, dx, \quad \forall \Theta \in V, \quad (3.23b)$$

for one $i \in \{1, \ldots, K_1\}$.

From Section 3.1.2, in particular (3.19) and (3.21), we know that then u solves (1.7) with right-hand side $f := f_i$ and $g := 0$. p is the solution of the corresponding adjoint equation (cf. (3.20)). In other words, u is then the deformation which arises if $\mathcal{O}$ is subjected only to <u>one</u> of the K_1 <u>deterministic</u> basis volume forces, namely f_i, and no surface load, i.e. $g := 0$; p is the corresponding adjoint state. Let us therefore denote u and p as $u^{(i,0)}$ and $p^{(i,0)}$, respectively. That way, we can obtain solutions $u^{(i,0)}$ and $p^{(i,0)}$ for each single basis volume force, i.e. for all $i = 1, \ldots, K_1$.

Similarly, if u and p satisfy

$$0 = \int_{\mathcal{O}} j'(u) \cdot \Theta \, dx + \int_{\partial \mathcal{O}} k'(u) \cdot \Theta \, ds + \int_{\mathcal{O}} Ae(\Theta) : e(p) \, dx, \quad \forall \Theta \in V, \quad (3.24a)$$

$$0 = \int_{\mathcal{O}} Ae(u) : e(\Theta) \, dx - \int_{\Gamma_N} g_j \cdot \Theta \, ds, \quad \forall \Theta \in V, \quad (3.24b)$$

for one $j \in \{1, \ldots, K_2\}$,

then we know again from (3.19) and (3.21) that these u and p are solutions to (1.7) with the right-hand side $f := 0$ and $g := g_j$, and to the corresponding adjoint equation (3.20), respectively. Let us refer to these solutions, that arise if the body $\mathcal{O}$ is subjected only to <u>one</u> of the K_2 <u>deterministic</u> surface loads, as $u^{(0,j)}$ and $p^{(0,j)}$, for $j \in \{1, \ldots, K_2\}$. Then we can easily construct a solution for a particular scenario $\omega \in \{\omega_1, \ldots, \omega_S\}$ as shown in the following Theorem.

---
[5] Note that these probabilitites are not connected with the probability estimates for the actual scenarios in any way.

3.2 Reformulation and Solution Plan for the Expectation-Based Model

Theorem 3.13. *For each $i \in \{1,\ldots,K_1\}$ and $j \in \{1,\ldots,K_2\}$, let $u^{(i,0)}, u^{(0,j)}, p^{(i,0)}$, and $p^{(0,j)}$ be given as described above[6]. Furthermore, let $\omega \in \{\omega_1,\ldots,\omega_S\}$ be a given scenario. Then*

$$\bar{u}(\mathscr{O};\omega) := \sum_{i=1}^{K_1} c_i^f(\omega) u^{(i,0)} + \sum_{j=1}^{K_2} c_j^g(\omega) u^{(0,j)} \tag{3.25}$$

is the solution to (3.1). A similar construction yields the adjoint state[7] for scenario ω: If the random coefficients $c_i^f(\omega)$ and $c_j^g(\omega)$ either satisfy

$$\sum_{i=1}^{K_1} c_i^f(\omega) + \sum_{j=1}^{K_2} c_j^g(\omega) = 1, \tag{3.26}$$

or

$$\sum_{i=1}^{K_1} c_i^f(\omega) + \sum_{j=1}^{K_2} c_j^g(\omega) = 0, \tag{3.27}$$

then there exist a $p_0 \in V$, which does not depend on ω, and a constant $C = C(\omega) \in \mathbb{R}$, $C \neq 0$, such that

$$\bar{p}(\mathscr{O};\omega) := \sum_{i=1}^{K_1} \frac{c_i^f(\omega)}{C} p^{(i,0)} + \sum_{j=1}^{K_2} \frac{c_j^g(\omega)}{C} p^{(0,j)} - p_0 \tag{3.28}$$

is the solution to the adjoint equation (3.20) belonging to the state $\bar{u}(\mathscr{O};\omega)$.

Proof. We need to check if (3.21) holds for $\bar{u}(\mathscr{O};\omega)$ and an arbitrary $\Theta \in V$. In doing so, we obtain by (3.22) and due to linearity that

$$\int_{\mathscr{O}} Ae(\bar{u}(\mathscr{O};\omega)) : e(\Theta)\,\mathrm{d}x - \int_{\mathscr{O}} f(\omega) \cdot \Theta\,\mathrm{d}x - \int_{\Gamma_N} g(\omega) \cdot \Theta\,\mathrm{d}s$$

$$= \sum_{i=1}^{K_1} c_i^f(\omega) \int_{\mathscr{O}} Ae\left(u^{(i,0)}\right) : e(\Theta)\,\mathrm{d}x + \sum_{j=1}^{K_2} c_j^g(\omega) \int_{\mathscr{O}} Ae\left(u^{(0,j)}\right) : e(\Theta)\,\mathrm{d}x$$

$$- \sum_{i=1}^{K_1} c_i^f(\omega) \int_{\mathscr{O}} f_i \cdot \Theta\,\mathrm{d}x - \sum_{j=1}^{K_2} c_j^g(\omega) \int_{\Gamma_N} g_j \cdot \Theta\,\mathrm{d}s$$

$$= \sum_{i=1}^{K_1} c_i^f(\omega) \left[\int_{\mathscr{O}} Ae\left(u^{(i,0)}\right) : e(\Theta)\,\mathrm{d}x - \int_{\mathscr{O}} f_i \cdot \Theta\,\mathrm{d}x \right]$$

---
[6] It is worth noticing that each of these solutions are obtained by solving elasticity PDEs for the basis forces — not the actual forces that constitute the scenarios.
[7] The adjoint state plays a role in the shape derivative, see Section 4.2.

$$+ \sum_{j=1}^{K_2} c_j^g(\omega) \left[\int_{\mathcal{O}} Ae\left(u^{(0,j)}\right) : e(\Theta) \, dx - \int_{\Gamma_N} g_j \cdot \Theta \, ds \right].$$

This last expression equals 0 because of (3.23b) and (3.24b). Therefore, $\bar{u}(\mathcal{O}; \omega)$ is indeed the solution to (3.1) with the forces $f(\omega)$ and $g(\omega)$.

To show that $\bar{p}(\mathcal{O}; \omega)$ is the solution to (3.20) belonging to the state $\bar{u}(\mathcal{O}; \omega)$, we have to check if (3.19) holds for all $\Theta \in V$. Contrary to the proof above, that $\bar{u}(\mathcal{O}; \omega)$ satisfies (3.21), there is a slight technical difference arising due to the occurrence of j' and k' in (3.19). Recall from Remark 3.3, that both $j(u)$ and $k(u)$ are assumed to be at most quadratic in u. As a consequence, $j'(u)$ and $k'(u)$ contain at most linear terms — but they might also contain constant ones. In other words, $j'(u)$ and $k'(u)$ might be *affine* in u, and hence the above argument concerning linearity does not apply directly here. We will see, that this issue is compensated by the additional term $-p_0$ in (3.28).

Now let $\Theta \in V$, and

$$j(u) = a_0 + a_1 \cdot u + a_2 u^2, \tag{3.29}$$

$$k(u) = b_0 + b_1 \cdot u + b_2 u^2, \tag{3.30}$$

with $a_0 \in L^2(\mathcal{O})$, $a_1 \in L^2(\mathcal{O}; \mathbb{R}^2)$, $a_2 \in L^\infty(\mathcal{O})$, $b_0 \in L^2(\partial \mathcal{O})$, $b_1 \in L^2(\partial \mathcal{O}; \mathbb{R}^2)$, and $b_2 \in L^\infty(\partial \mathcal{O})$. Then

$$j'(u) = a_1 + 2a_2 u, \tag{3.31}$$

$$k'(u) = b_1 + 2b_2 u. \tag{3.32}$$

Moreover, let $C := \sum_{i=1}^{K_1} c_i^f(\omega) + \sum_{j=1}^{K_2} c_j^g(\omega)$. We distinguish two cases:

Case 1: $C \neq 0$

Note that because of the two prerequisites (3.26) and (3.27), this means that $C = 1$. However, we keep writing C for a while to demonstrate the problems that may arise in general if no such conditions are satisfied, and $a_2 \neq 0$ as well as $b_2 \neq 0$. In this case, we may divide by C, yielding

$$\sum_{i1}^{K_1} \frac{c_i^f(\omega)}{C} + \sum_{j=1}^{K_2} \frac{c_j^g(\omega)}{C} = 1. \tag{3.33}$$

Next, we plug $\bar{u}(\mathcal{O}; \omega)$, $\bar{p}(\mathcal{O}; \omega)$, (3.31), and (3.32) into (3.19) and obtain

$$\int_{\mathcal{O}} j'(\bar{u}(\mathcal{O}; \omega)) \cdot \Theta \, dx + \int_{\partial \mathcal{O}} k'(\bar{u}(\mathcal{O}; \omega)) \cdot \Theta \, ds + \int_{\mathcal{O}} Ae(\Theta) : e(\bar{p}(\mathcal{O}; \omega)) \, dx$$

3.2 Reformulation and Solution Plan for the Expectation-Based Model

$$= \sum_{i=1}^{K_1} \frac{c_i^f(\omega)}{C} \int_{\mathscr{O}} j'\left(u^{(i,0)}\right) \cdot \Theta \, dx + \sum_{j=1}^{K_2} \frac{c_j^g(\omega)}{C} \int_{\mathscr{O}} j'\left(u^{(0,j)}\right) \cdot \Theta \, dx$$

$$+ \int_{\mathscr{O}} 2a_2(C-1) \left[\sum_{i=1}^{K_1} \frac{c_i^f(\omega)}{C} u^{(i,0)} + \sum_{j=1}^{K_2} \frac{c_j^g(\omega)}{C} u^{(0,j)} \right] \cdot \Theta \, dx \quad (3.34)$$

$$+ \sum_{i=1}^{K_1} \frac{c_i^f(\omega)}{C} \int_{\partial \mathscr{O}} k'\left(u^{(i,0)}\right) \cdot \Theta \, ds + \sum_{j=1}^{K_2} \frac{c_j^g(\omega)}{C} \int_{\partial \mathscr{O}} k'\left(u^{(0,j)}\right) \cdot \Theta \, ds$$

$$+ \int_{\partial \mathscr{O}} 2b_2(C-1) \left[\sum_{i=1}^{K_1} \frac{c_i^f(\omega)}{C} u^{(i,0)} + \sum_{j=1}^{K_2} \frac{c_j^g(\omega)}{C} u^{(0,j)} \right] \cdot \Theta \, ds \quad (3.35)$$

$$+ \sum_{i=1}^{K_1} \frac{c_i^f(\omega)}{C} \int_{\mathscr{O}} Ae(\Theta) : e\left(p^{(i,0)}\right) dx + \sum_{j=1}^{K_2} \frac{c_j^g(\omega)}{C} \int_{\mathscr{O}} Ae(\Theta) : e\left(p^{(0,j)}\right) dx$$

$$- \int_{\mathscr{O}} Ae(\Theta) : e(p_0) \, dx.$$

Note that the above equality holds in particular because of (3.33), which can be seen exemplarily[8] for j' as follows:

$$j'(\bar{u}(\mathscr{O};\omega)) = 1a_1 + 2a_2 \left(\sum_{i=1}^{K_1} c_i^f(\omega) u^{(i,0)} + \sum_{j=1}^{K_2} c_j^g(\omega) u^{(0,j)} \right)$$

$$= \left(\sum_{i=1}^{K_1} \frac{c_i^f(\omega)}{C} + \sum_{j=1}^{K_2} \frac{c_j^g(\omega)}{C} \right) a_1$$

$$+ 2a_2 \left(\sum_{i=1}^{K_1} c_i^f(\omega) u^{(i,0)} + \sum_{j=1}^{K_2} c_j^g(\omega) u^{(0,j)} \right)$$

$$= \sum_{i=1}^{K_1} \frac{c_i^f(\omega)}{C} \left[a_1 + 2a_2 C u^{(i,0)} \right] + \sum_{j=1}^{K_2} \frac{c_j^g(\omega)}{C} \left[a_1 + 2a_2 C u^{(0,j)} \right]$$

$$= \sum_{i=1}^{K_1} \frac{c_i^f(\omega)}{C} j'\left(u^{(i,0)}\right) + \sum_{j=1}^{K_2} \frac{c_j^g(\omega)}{C} j'\left(u^{(0,j)}\right)$$

$$+ 2a_2(C-1) \underbrace{\left[\sum_{i=1}^{K_1} \frac{c_i^f(\omega)}{C} u^{(i,0)} + \sum_{j=1}^{K_2} \frac{c_j^g(\omega)}{C} u^{(0,j)} \right]}_{=\frac{1}{C} \bar{u}(\mathscr{O};\omega)}.$$

---

[8] The same can be done for k' almost verbatim, only the notations from (3.32) have to be used instead.

At this point, we make use of requirement (3.26) which tells us that $C = 1$. We see that both terms, (3.34) and (3.35), vanish. Now we can choose $p_0 := 0$, and the above computations show, taking (3.23a) and (3.24a) into account, that (3.19) holds as desired. Note on the other hand that if C were $C \neq 1$, $C \neq 0$, and $a_2 \neq 0$ or $b_2 \neq 0$, (3.34) and (3.35) would not vanish, and p_0 would have to be chosen such that $-\int_{\mathscr{O}} Ae(\Theta) : e(p_0)\, dx$ cancels these terms out. It is, of course, possible to find[9] such a p_0, but since the non-vanishing terms contain $\bar{u}(\mathscr{O};\omega)$, p_0 would depend on ω. This is clearly not desirable from a computational point of view. See Remark 3.14 for further remarks and observations concerning this issue.

Case 2: $C = 0$

This case corresponds to requirement (3.27). At first, we note that exemplarily j' from (3.31) can be rewritten[10] using (3.25) as

$$j'(\bar{u}(\mathscr{O};\omega)) = a_1 + 0 + 2a_2 \left(\sum_{i=1}^{K_1} c_i^f(\omega) u^{(i,0)} + \sum_{j=1}^{K_2} c_j^g(\omega) u^{(0,j)} \right)$$

$$= a_1 + \left(\sum_{i=1}^{K_1} c_i^f(\omega) + \sum_{j=1}^{K_2} c_j^g(\omega) \right) a_1$$

$$+ 2a_2 \left(\sum_{i=1}^{K_1} c_i^f(\omega) u^{(i,0)} + \sum_{j=1}^{K_2} c_j^g(\omega) u^{(0,j)} \right)$$

$$= a_1 + \sum_{i=1}^{K_1} c_i^f(\omega) j'\left(u^{(i,0)}\right) + \sum_{j=1}^{K_2} c_j^g(\omega) j'\left(u^{(0,j)}\right).$$

With this in mind, we set $C := 1$ in (3.28) and check if (3.19) is satisfied:

$$\int_{\mathscr{O}} j'(\bar{u}(\mathscr{O};\omega)) \cdot \Theta\, dx + \int_{\partial \mathscr{O}} k'(\bar{u}(\mathscr{O};\omega)) \cdot \Theta\, ds + \int_{\mathscr{O}} Ae(\Theta) : e(\bar{p}(\mathscr{O};\omega))\, dx$$

$$= \sum_{i=1}^{K_1} c_i^f(\omega) \int_{\mathscr{O}} j'\left(u^{(i,0)}\right) \cdot \Theta\, dx + \sum_{j=1}^{K_2} c_j^g(\omega) \int_{\mathscr{O}} j'\left(u^{(0,j)}\right) \cdot \Theta\, dx$$

$$+ \sum_{i=1}^{K_1} c_i^f(\omega) \int_{\partial \mathscr{O}} k'\left(u^{(i,0)}\right) \cdot \Theta\, ds + \sum_{j=1}^{K_2} c_j^g(\omega) \int_{\partial \mathscr{O}} k'\left(u^{(0,j)}\right) \cdot \Theta\, ds$$

$$+ \sum_{i=1}^{K_1} c_i^f(\omega) \int_{\mathscr{O}} Ae(\Theta) : e\left(p^{(i,0)}\right)\, dx + \sum_{j=1}^{K_2} c_j^g(\omega) \int_{\mathscr{O}} Ae(\Theta) : e\left(p^{(0,j)}\right)\, dx$$

---

[9] p_0 is just a solution of a PDE.
[10] The same can be done for k' from (3.32).

3.2 Reformulation and Solution Plan for the Expectation-Based Model

$$+ \int_{\mathcal{O}} a_1 \cdot \Theta \, dx + \int_{\partial \mathcal{O}} b_1 \cdot \Theta \, ds - \int_{\mathcal{O}} Ae(\Theta) : e(p_0) \, dx.$$

Next, choose $p_0 \in V$ such that

$$\int_{\mathcal{O}} Ae(\Theta) : e(p_0) \, dx = \int_{\mathcal{O}} a_1 \cdot \Theta \, dx + \int_{\partial \mathcal{O}} b_1 \cdot \Theta \, ds, \; \forall \Theta \in V. \tag{3.36}$$

Comparing (3.36) to (1.11) reveals that (3.36) is also an elasticity PDE. Because of the assumptions on a_1 and b_1, we can apply the Lax-Milgram Theorem A.6 to (3.36), which shows the (unique) existence of such a p_0. Most importantly, as the right-hand side in (3.36) is independent of ω, p_0 also does not depend on ω.

Knowing this, we see together with the above computations, (3.23a), and (3.24a) that $\bar{p}(\mathcal{O}; \omega)$ indeed satisfies condition (3.19). □

Remark 3.14. 1. *The proof of Theorem (3.13) showed that the constant $C = C(\omega)$ in (3.28) is always 1. The only reason that we included this constant in the formulation of the theorem is to emphasize the difficulties that might arise if neither (3.26) nor (3.27) hold.*

2. *The requirement (3.26), that all random coefficients must add up to 1, can always be satisfied by choosing the basis forces $f_1, \ldots, f_{K_1}$ and $g_1, \ldots, g_{K_2}$ appropriately. For instance, in two dimensions three basis forces suffice to be able to combine them linearly to obtain any given force in such a way that (3.26) is satisfied in addition. Note however, that simply rescaling of the basis forces would not yield the desired property (3.26) because the rescaling factors would depend on the scenario ω, which yields, therefore, basis forces that are not deterministic any more — a crucial property of the basis forces for the following reason: Theorem 3.13 tells us that it suffices to solve as many elasticity PDEs as there are basis forces, i.e. $K_1 + K_2$, to obtain the solution $\bar{u}(\mathcal{O}; \omega)$ for any scenario ω. Moreover, at most $K_1 + K_2 + 1$ elasticity PDEs[11] have to be solved to be able to construct the adjoint states $\bar{p}(\mathcal{O}; \omega)$ for all scenarios ω. So the number of PDEs that need to be solved is independent of the number of scenarios S, provided the basis forces are deterministic. This fact constitutes the algorithmic shortcut we have mentioned already in the beginning of this chapter, as the solutions of the elasticity PDEs are the most time consuming parts in the computation.*

---

[11] The number of PDEs to be solved to get the adjoint state $\bar{p}(\mathcal{O}; \omega)$ is at most $K_1 + K_2 + 1$; the "+1" is arising from (3.36) to obtain p_0 in that case.

3. *In case of the compliance objective functional (see* (1.14) *and Remark 3.10), Theorem 3.13 can be simplified. In particular, we then have that* $j(u) = f(\omega) \cdot u$ *and* $k(u) = (g(\omega) \cdot u) \chi_{\Gamma_N}$. *Condition* (3.26) *is not necessary in this case, the sum of all random coefficients can be anything, as* $a_2 = b_2 = 0$ *in* (3.29) *and* (3.30), *respectively. Additionally,* $a_1 = f(\omega)$ *and* $b_1 = g(\omega)$ *which means because of* (3.22) *and* (3.25), *that those terms that are constant with respect to u in* (3.31) *and* (3.32) *can be decomposed precisely the same way as* $\bar{u}(\mathscr{O}; \omega)$. *This is why even though* j' *and* k' *are constant with respect to u, the check of condition* (3.19) *in the proof of Theorem 3.13 goes through as if* j' *and* k' *were linear. As a consequence,* (3.28) *simplifies in this case for arbitrary random coefficients to*

$$\bar{p}(\mathscr{O}; \omega) = \sum_{i=1}^{K_1} c_i^f(\omega) p^{(i,0)} + \sum_{j=1}^{K_2} c_j^g(\omega) p^{(0,j)}.$$

4. *In case of the quadratic objective functional* (1.15), *Theorem 3.13 might be simplified, depending on* u_0. *We have in particular in this case*

$$j(u) = (u - u_0)^2 = u_0^2 - 2u_0 \cdot u + u^2, \quad k(u) \equiv 0.$$

Now there are two possible cases:

Case 1: $u_0 = 0$ *Then,* $j(u) = u^2$, *and consequently* $j'(u) = 2u$. *We see that* j' *is linear in u. Because of that, condition* (3.26) *can be relaxed, and the proof of Theorem 3.13 shows that* p_0 *can be chosen* $p_0 := 0$, *and* $C := 1$ *in* (3.28).

Case 2: $u_0 \neq 0$ *In this case, we have that* $j'(u) = -2u_0 + 2u$. *This means that* a_1 *and* a_2 *in* (3.31) *are not 0. Therefore, the theorem cannot be simplified, i.e. either* (3.26), *or* (3.27) *must be satisfied and* p_0 *must be computed according to* (3.36).

Theorem 3.13 enables us to rewrite our two-stage stochastic shape optimization problem (3.7) in a slightly different way which highlights the idea of the algorithmic shortcut due to the approach with the deterministic basis forces. This reformulation is given in the next Corollary.

3.2 Reformulation and Solution Plan for the Expectation-Based Model

Corollary 3.15. *Problem (3.7) is equivalent to the following minimization problem:*

$$\min\left\{ \begin{array}{l} \alpha \int_{\mathcal{O}} 1\, dx + \beta \int_{\partial \mathcal{O}} 1\, ds \\ + \sum_{\mu=1}^{S} \pi_{\mu} \left(\int_{\mathcal{O}} j\left(x, \bar{u}(\mathcal{O}; \omega_{\mu})\right) dx + \int_{\partial \mathcal{O}} k\left(x, \bar{u}(\mathcal{O}; \omega_{\mu})\right) ds \right) : \\ \mathcal{O} \in \mathcal{U}_{ad}, \\ \bar{u}(\mathcal{O}; \omega_{\mu}) := \sum_{i=1}^{K_1} c_i^f(\omega_{\mu}) u^{(i,0)} + \sum_{j=1}^{K_2} c_j^g(\omega_{\mu}) u^{(0,j)},\ \mu = 1, \ldots, S \end{array} \right\}. \quad (3.37)$$

Proof. This follows immediately from (3.7) and Theorem 3.13. □

From (3.37) we can see how the objective value can be evaluated quite easily for a given shape $\mathcal{O}$. We emphasize that in order to compute $\bar{u}(\mathcal{O}; \omega)$, we only need to solve $K_1 + K_2$ elasticity PDEs according to (3.25). The big advantage in that approach is that this number of PDEs is independent of the actual number of scenarios S — for different scenarios, the solutions $u^{(i,0)}$ and $u^{(0,j)}$ corresponding to the basis forces just have to be combined linearly with the corresponding factors. The next step towards the solution of (3.37) using a steepest descent algorithm would be the computation of a descent direction[12]. Then we obtain new shapes in every iteration, and evaluate the objective function for this new iterate according to formulation (3.37). This procedure constitutes the solution plan, namely in each iteration the objective function is evaluated for a given shape $\mathcal{O}$ according to these steps[13]:

Algorithm 3.16 (Evaluation of the objective functional of (3.37)).

1. Solve (3.23b) for all $i = 1, \ldots, K_1$, to obtain $u^{(i,0)}$ for $i = 1, \ldots, K_1$, and solve (3.24b) for all $j = 1, \ldots, K_2$, for $u^{(0,j)}$, $j = 1, \ldots, K_2$. This amounts to the solution of $K_1 + K_2$ elasticity PDEs employing composite finite elements as described in Chapter 2.

2. If the objective function is not the compliance and the adjoint states are needed for the shape derivative, solve (3.23a) for all of the just computed $u^{(i,0)}$ and $u^{(0,j)}$, $i = 1, \ldots, K_1$, $j = 1, \ldots, K_2$, to obtain $p^{(i,0)}$ and $p^{(0,j)}$ for all $i = 1, \ldots, K_1$, $j = 1, \ldots, K_2$. Depending on the objective function and the random coefficients, one might have to solve (3.36) additionally for p_0. This amounts to the solution of $K_1 + K_2$ or $K_1 + K_2 + 1$ elasticity PDEs, respectively.

---

[12] Cf. Section 4.2
[13] We assume that the basis forces $f_1, \ldots, f_{K_1}$, $g_1, \ldots, g_{K_2}$ and the random coefficients $c_i^f(\omega), i = 1, \ldots, K_1$, and $c_j^g(\omega), j = 1, \ldots, K_2$, are chosen such that either (3.26) or (3.27) holds. This is not necessary in some special cases, see Remark 3.14.

3. Assemble $\bar{u}(\mathcal{O};\omega)$ according to (3.25) for each scenario $\omega \in \{\omega_1,\ldots,\omega_S\}$, and compute the objective functional as stated in formulation (3.37). If required, assemble the adjoint state $\bar{p}(\mathcal{O};\omega)$ according to (3.28).

Not only does the above algorithm compute the necessary ingredients needed to evaluate the objective functional in (3.37), but in addition it also computes the adjoint states for all scenarios if they are required. These will be needed for the shape derivative and hence for the descent direction anyway.

3.3 Expected Excess

In Section 3.1.1 we introduced the two-stage stochastic shape optimization problem

$$\min\left\{\bar{G}(\mathcal{O};\omega) : \mathcal{O} \in \mathcal{U}_{ad}\right\}.$$

We then considered at first the expectation-based model (3.7) where the random variables $\bar{G}(\mathcal{O};\omega)$ are ranked according to their expectation values. In this section, we introduce the model involving the *expected excess* as risk measure (cf. Definition 1.12 on page 22). Additionally, we introduce two formulations that give rise to two different ways to solve this problem. The shape optimization problems arising from these formulations can then be solved employing the techniques described in Chapter 4.

Definition 3.17. *Analogously to Section 1.3.2, we obtain the random shape optimization problem with the expected excess risk measure and the preselected tolerance level $\eta \in \mathbb{R}$ as:*

$$\min\left\{\mathcal{Q}_{EE_\eta}\left(\bar{G}(\mathcal{O};\omega)\right) : \mathcal{O} \in \mathcal{U}_{ad}\right\}, \tag{3.38}$$

where

$$\begin{aligned}\mathcal{Q}_{EE_\eta}\left(\bar{G}(\mathcal{O};\omega)\right) &= \mathbb{E}\left(\max\left\{\bar{G}(\mathcal{O};\omega) - \eta, 0\right\}\right) \\ &= \sum_{\mu=1}^{S} \pi_\mu \max\left\{\bar{G}(\mathcal{O};\omega_\mu) - \eta, 0\right\} \\ &= \sum_{\mu=1}^{S} \pi_\mu \max\left\{\mathbf{J}(\mathcal{O};\omega_\mu) - \eta, 0\right\}.\end{aligned}$$

In what follows, we describe two approaches, based on two different ideas, to solve problem (3.38) numerically.

3.3.1 Barrier Method

What prevents us from applying a steepest descent method directly to problem (3.38) is the max-expression in its objective, because it is not differentiable. One way to cope with this, is to follow the idea from the finite dimensional linear case: With Lemma 1.13 on page 23 in mind, we introduce additional variables $t_\mu, \mu = 1, \ldots, S$, and rewrite problem (3.38) equivalently as follows:

$$\min\left\{\sum_{\mu=1}^{S} \pi_\mu t_\mu : \begin{array}{l} \mathbf{J}(\mathcal{O};\omega_\mu) - \eta \leq t_\mu, \mu = 1, \ldots, S, \\ 0 \leq t_\mu, \mu = 1, \ldots, S, \\ \mathcal{O} \in \mathcal{U}_{\mathrm{ad}} \end{array}\right\}. \tag{3.39}$$

After this first step, the shape functional $\mathbf{J}(\mathcal{O};\omega)$ now appears in inequality constraints. In the second step, we eliminate these inequality constraints by considering an approximate problem

$$\min\{\mathcal{B}(t, \mathcal{O}; \gamma) : \mathcal{O} \in \mathcal{U}_{\mathrm{ad}}\}, \tag{3.40}$$

with

$$\mathcal{B}(t, \mathcal{O}; \gamma) := \sum_{\mu=1}^{S} \pi_\mu t_\mu - \gamma \left(\sum_{\mu=1}^{S} \ln\left(-\mathbf{J}(\mathcal{O};\omega_\mu) + \eta + t_\mu\right) + \sum_{\mu=1}^{S} \ln\left(t_\mu\right) \right), \tag{3.41}$$

for $t \in \mathbb{R}^S, t > 0$, and $\mathcal{O} \in \mathcal{U}_{\mathrm{ad}}$. This procedure amounts to the classical *barrier method* (cf. [NW99, Rus06, GK02, BGLS03, Ye97]), which is an interior point method with the following basic idea:

In problem (3.40), γ is a positive parameter, also referred to as *barrier parameter*. Suppose, we have $\mathcal{O} \in \mathcal{U}_{\mathrm{ad}}$ and $t \in \mathbb{R}^S$ feasible for problem (3.39) such that[14]

$$\mathbf{J}(\mathcal{O};\omega_\mu) - \eta < t_\mu \text{ and } 0 < t_\mu, \forall \mu = 1, \ldots, S.$$

Particularly, this means that $-\mathbf{J}(\mathcal{O};\omega_\mu) + \eta + t_\mu > 0$ for all $\mu = 1, \ldots, S$, hence all terms occurring in $\mathcal{B}(t, \mathcal{O}; \gamma)$ are well defined. For small but still positive values of $-\mathbf{J}(\mathcal{O};\omega_\mu) + \eta + t_\mu$ and t_μ, the functions $-\gamma \ln\left(-\mathbf{J}(\mathcal{O};\omega_\mu) + \eta + t_\mu\right)$ and $-\gamma \ln(t_\mu)$ create a "barrier", preventing $-\mathbf{J}(\mathcal{O};\omega_\mu) + \eta + t_\mu$ and t_μ from becoming too close to 0. We then let the barrier parameter γ tend to 0, and under certain conditions[15] the solution to problem (3.40) approaches a solution to problem (3.39).

---

[14] Note that such $\mathcal{O} \in \mathcal{U}_{\mathrm{ad}}$ and $t \in \mathbb{R}^S, t > 0$, can always be found, the components of t just have to be chosen big enough. In other words, this means we have an interior point.

[15] At least in the case of finite dimensional nonlinear optimization problems.

For the specific shape optimization problem (3.39) we have at hand here, it is difficult to obtain convergence results, starting with the issue of existence of optimal solutions to problems (3.39) and (3.40), which we have mentioned already in Chapter 1. However, following the basic idea of barrier methods as described above, we at least obtain a heuristic solution method to tackle problem (3.38): For decreasing barrier parameters γ, we solve a sequence of approximate problems (3.40) using the steepest descent method described in Chapter 4. The descent direction in this case does not only involve the shape derivative, i.e. the derivative of $\mathscr{B}$ with respect to $\mathscr{O}$, but also the derivative of $\mathscr{B}$ with respect to t.

3.3.2 Smooth Approximation

In this section, we describe an idea different from the barrier approach in Section 3.3.1 to solve problem (3.38). In the setting of finite dimensional linear programs, introducing additional variables according to Lemma 1.13 makes sense and is advantageous, as one gets rid of the max-expression in the objective function and obtains a linear program, which all the theory and techniques known from linear programming can be applied to. Our particular shape optimization problem (3.38), however, is a nonlinear problem, and the problem we obtain by following the ideas from Lemma 1.13, i.e. problem (3.39), still is in the class of nonlinear problems. Here we suggest, instead of adding more variables, to simply approximate the max-expression in (3.38) smoothly, and solve the arising approximate problem using the techniques from Chapter 4.

The idea is based on the following observation: For any $a \in \mathbb{R}$,

$$\max\{a,0\} = \frac{|a|+a}{2} = \frac{\sqrt{a^2}+a}{2},$$

which we approximate for a small $\varepsilon > 0$ by

$$\frac{\sqrt{a^2+\varepsilon}+a}{2}.$$

Using this, we simply replace problem (3.38) by the approximate problem

$$\min\left\{\sum_{\mu=1}^{S} \pi_\mu \frac{\sqrt{(\mathbf{J}(\mathscr{O};\omega_\mu)-\eta)^2+\varepsilon}+(\mathbf{J}(\mathscr{O};\omega_\mu)-\eta)}{2} : \mathscr{O} \in \mathscr{U}_{\mathrm{ad}}\right\}. \quad (3.42)$$

3.4 Excess Probability

Now we focus on the random shape optimization problem where the random variables $\bar{G}(\mathcal{O};\omega)$ are ranked according to the *excess probability* risk measure $\mathcal{Q}_{\text{EP}_\eta}$, defined in Definition 1.12 on page 22. First, we give its definition, and then we introduce an approximate problem that can be solved by the method from Chapter 4.

Definition 3.18. *Let $\eta \in \mathbb{R}$ be a preselected tolerance threshold. Then, analogously to Section 1.3.2, the random shape optimization problem with the excess probability risk measure is given as*

$$\min\left\{\mathcal{Q}_{\text{EP}_\eta}\left(\bar{G}(\mathcal{O};\omega)\right) : \mathcal{O} \in \mathcal{U}_{ad}\right\}, \tag{3.43}$$

where

$$\mathcal{Q}_{\text{EP}_\eta}\left(\bar{G}(\mathcal{O};\omega)\right) = \mathbb{P}\left(\{\omega \in \Omega : \bar{G}(\mathcal{O};\omega) > \eta\}\right)$$
$$= \mathbb{P}\left(\{\omega \in \Omega : \mathbf{J}(\mathcal{O};\omega) > \eta\}\right).$$

Since we assumed that there are finitely many scenarios $\omega_i, i = 1,\ldots,S$, occurring with probabilities $\pi_i, i = 1,\ldots,S$, we can express the probability in problem (3.43) as

$$\mathbb{P}(\{\omega \in \Omega : \mathbf{J}(\mathcal{O};\omega) > \eta\}) = \sum_{i=1}^{S} \pi_i \mathcal{H}\left(\mathbf{J}(\mathcal{O};\omega_i) - \eta\right),$$

where $\mathcal{H}(x) := \begin{cases} 0 & x \leq 0, \\ 1 & x > 0, \end{cases}$ for $x \in \mathbb{R}$, denotes the Heaviside function. The idea is now to use a smooth approximation of $\mathcal{H}(x)$, such as

$$\mathcal{H}(x) \approx \frac{1}{2} + \frac{1}{2}\tanh(\kappa x)$$
$$= \frac{1}{2} + \frac{1}{2}\frac{\sinh(\kappa x)}{\cosh(\kappa x)}$$
$$= \frac{1}{2} + \frac{1}{2}\frac{\frac{e^{\kappa x} - e^{-\kappa x}}{2}}{\frac{e^{\kappa x} + e^{-\kappa x}}{2}}$$
$$= \frac{1}{2} + \frac{1}{2}\frac{e^{2\kappa x} - 1}{e^{2\kappa x} + 1}$$
$$= \frac{e^{2\kappa x} + 1 + e^{2\kappa x} - 1}{2(e^{2\kappa x} + 1)}$$

$$= \frac{1}{1+e^{-2\kappa x}}.$$

Larger values for κ result in sharper transitions at $x = 0$. If we define $\mathscr{H}(0) := \frac{1}{2}$, we get equality in the limit:

$$\mathscr{H}(x) = \lim_{\kappa \to \infty} \frac{1}{2}(1 + \tanh(\kappa x)) = \lim_{\kappa \to \infty} \frac{1}{1+e^{-2\kappa x}}.$$

In Fig. 3.1, we plotted the above approximation of $\mathscr{H}$ for different values of κ.

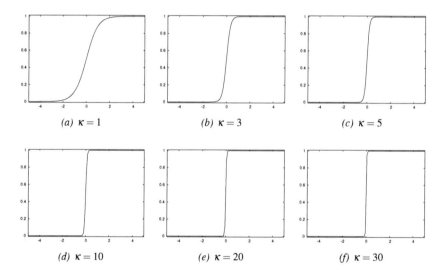

Fig. 3.1: The approximation for $\mathscr{H}(x)$ is shown for different values for κ.

Putting it all together, we obtain the following approximate problem to problem (3.43):

$$\min\left\{ \sum_{i=1}^{S} \pi_i \frac{1}{1+e^{-2\kappa(\mathbf{J}(\mathscr{O};\omega_i)-\eta)}} : \mathscr{O} \in \mathscr{U}_{\mathrm{ad}} \right\}. \tag{3.44}$$

3.4 Excess Probability

Remark 3.19. *We assume that the random forces in problems* (3.40), (3.42), *and* (3.44) *also have the structure described in Definition 3.11. Then the evaluation of* $\mathbf{J}(\mathcal{O};\omega)$ *for a given shape* $\mathcal{O} \in \mathcal{U}_{ad}$ *and a scenario* $\omega \in \Omega$ *is also done analogously to Algorithm 3.16. In particular, only elasticity PDEs for the basis forces have to be solved.*

4 Solving Shape Optimization Problems

This chapter is dedicated to the actual numerical solution techniques we implemented to solve the (random) shape optimization problems described in Chapter 3. As noted in the beginning, we employed a steepest descent algorithm (see Section 4.4) together with a level set method (see Section 4.1). The necessary function evaluations are done according to Algorithm 3.16, whereas the computation of the descent direction is described here in this chapter, making use of the shape derivative (see Section 4.2) and also the topological derivative (see Section 4.3).

There are various methods that aim to solve shape optimization problems, and before we start describing our particular level set approach, we briefly mention some of these methods. For example, there is the *homogenization method* (cf. Allaire [All02]) whose physical idea in principle consists of averaging heterogeneous media in order to derive effective properties. In [All02, Chapter 4], the method is applied to optimal design problems with linear elasticity in form of two-phase optimization problems. The task is then to find an optimal distribution of two elastic materials, i.e. there are no void areas. This results in an ill-posed optimization problem, which, however, homogenization theory provides a relaxation to by introducing generalized designs. Numerical examples can also be found in [HN97].

Another approach, namely topology optimization by the *material distribution method*, is described in the book by Bendsøe and Sigmund [BS03]. Each point in the design can have material or not[1]. In a discrete setting, there is a grid where each grid cell, or "pixel", is either filled with material, or there is none. This leads to nonlinear optimization problems with binary variables which indicate the presence or absence of material in the grid cells, respectively. In [SS03] for example, they show that certain nonlinear 0-1 topology optimization problems can be equivalently formulated as *linear* mixed 0-1 programs, which can be solved as such — at least on quite coarse grids. The idea described in [BS03], however, is to replace the integer variables with continuous ones, resulting in a density function with values between 0 and 1, and then to penalize intermediate values. This yields the so-called SIMP-model[2]. Various solution methods are mentioned in [BS03].

---

[1] Similar to a black and white image.
[2] Solid Isotropic Material with Penalization. Also see the website http://www.topopt.dtu.dk/, where some problems can be set up and solved online.

Claudia Stangl implemented this model in her diploma thesis [Sta08], also incorporating stochastic forces for the expectation-based problem, and solved it using IPOPT (cf. [WB06]). Maar and Schulz [MS00] describe the application of an interior point multigrid method for this type of problem.

Newton's method, involving second order shape derivatives (cf. [NR]), has been applied to some shape optimization problem for example in [NP02].

Level set methods provide another approach to tackling shape optimization problems. This is the method we applied to our problems, so we will describe it in more detail in the following section.

4.1 Level Set Formulation

The *level set method* provides a general framework for interface propagation using implicit representations. It was first introduced by Osher and Sethian [OS88], and general overviews can be found in [OF03, Set01]. The level set method has been applied to shape optimization problems for instance by Allaire et al. in [AJT04, AJ05] and Pach [Pac05]. Additionally, level set based shape optimization has been combined with the topological derivative in order to also optimize the number and shapes of holes in the design (cf. [AdGJT05, BHR04, dGAJ06, HKO07, FLSS07]). For an overview on level set methods for shape optimization problems and suitable descent methods we refer to [Bur03, BO05].

Definition 4.1. *A level set function ϕ is a (Lipschitz) continuous function defined on the whole working domain D. A domain $\mathscr{O} \subset D$ is identified with the level set function ϕ via the following definition:*

$$\begin{cases} \phi(x) = 0 & \Longleftrightarrow x \in D \cap \partial \mathscr{O}, \\ \phi(x) < 0 & \Longleftrightarrow x \in \mathscr{O}, \\ \phi(x) > 0 & \Longleftrightarrow x \in D \setminus \overline{\mathscr{O}}. \end{cases} \quad (4.1)$$

The concept of level set functions is illustrated in Fig. 4.1.

Obviously, there are infinitely many choices for level set functions. In practice, *signed distance functions* (cf. [OF03, Chapter 2]) are preferred for stability reasons in numerical computations. Signed distance functions are implicit functions ϕ with $|\phi(x)| = \text{dist}(\mathscr{O}, x)$ for all $x \in \mathbb{R}^2$, such that

- $\phi(x) = \text{dist}(\partial \mathscr{O}, x) = 0$ for all $x \in \partial \mathscr{O}$,
- $\phi(x) = -\text{dist}(\partial \mathscr{O}, x)$ for all $x \in \mathscr{O}$, and

4.1 Level Set Formulation

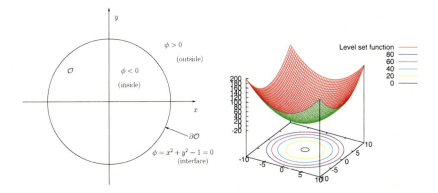

Fig. 4.1: On the left, the domain $\mathcal{O}$ is the unit disc, represented by the level set function $\phi(x,y) = x^2 + y^2 - 1$, i.e. $\mathcal{O} = \{(x,y) \in \mathbb{R}^2 : \phi(x,y) < 0\}$. On the right, we have plotted ϕ together with the contours corresponding to different levels of ϕ at the bottom.

- $\phi(x) = \text{dist}(\partial \mathcal{O}, x)$ for all $x \in \mathbb{R}^2 \setminus \overline{\mathcal{O}}$.

They have the additional property that $\|\nabla \phi\| = 1$ — except, of course, at points that have the same distance to at least two points on the interface, where $\nabla \phi$ fails to exist[3]. Given an initial contour, a signed distance function can be efficiently computed using the *fast marching method* (cf. [AS99, OF03]). In our numerical experiments, we also started the descent algorithm with a signed distance function, and occasionally reinitialized it to prevent the level set function from becoming too flat or too steep (also see [AJT04]).

One of the advantages of level set methods is the fact that the unit outward normal n can be simply expressed by means of the level set function ϕ (cf. [OF03]) as

$$n = \frac{\nabla \phi}{\|\nabla \phi\|}, \qquad (4.2)$$

and the mean curvature h as

$$h = \text{div}(n) = \text{div}\left(\frac{\nabla \phi}{\|\nabla \phi\|}\right). \qquad (4.3)$$

That way, n and h can be evaluated everywhere in D as ϕ is defined on D.

---

[3] Note that, as mentioned in [OF03], this is why the equation $\|\nabla \phi\| = 1$ is only true in a "general sense". However, such relations are still useful in numerical approximations.

Now suppose that the shape $\mathscr{O}$ evolves in fictitious time $t \geq 0$ with velocity V in normal direction[4]. Then we get a time-dependent domain $\mathscr{O}(t)$ represented by the level set function $\phi(t,x(t))$ such that $\dot{x}(t) = Vn$ and $x(0) = x \in \mathscr{O}$, and it holds that

$$\phi(t,x(t)) = 0, \quad \forall x(t) \in \partial \mathscr{O}(t).$$

Differentiating the above equation with respect to t and making use of the above relations yield the so-called *level set equation*

$$\begin{aligned} 0 &= \phi_t + \dot{x}(t) \cdot \nabla \phi \\ &= \phi_t + Vn \cdot \nabla \phi \\ &= \phi_t + V \frac{\nabla \phi}{\|\nabla \phi\|} \cdot \nabla \phi \\ &= \phi_t + V \|\nabla \phi\|. \end{aligned}$$

Therefore, we obtain the relation

$$V = -\frac{\phi_t}{\|\nabla \phi\|}, \tag{4.4}$$

which identifies the variation of the level set function, ϕ_t, with the variation of the level sets in normal direction, V. (4.4) will be useful later when rewriting the shape functionals and their derivatives in the level set context.

4.1.1 Computation of the Mean Curvature

Here we briefly describe the numerical computation of the mean curvature h in the level set context which will appear in the expression for the shape derivative in Section 4.2. We make use of expression (4.3). Recall from Chapter 2 that we use linear finite element basis functions, which is why directly evaluating expression (4.3) would not make any sense: (4.3) contains second derivatives of ϕ which would all be identically 0 because of the linear ansatz. Instead, we use a standard technique as described below.

Suppose, our grid consists of $N \in \mathbb{N}$ nodes, and we are interested in the values of h at each of these grid nodes. We will see that these values can be obtained by solving a system of linear equations. Therefore, let $b_1, \ldots, b_N$ be the standard (linear) finite element basis functions (see the beginning of Chapter 2) that vanish

---

[4] Note that variations in tangential direction do not change the level sets.

on ∂D. Keeping Notation 2.4 on page 30 in mind, we can make the following ansatz for h and ϕ in the standard finite element space:

$$h = \sum_{j=1}^{N} H_j b_j, \text{ and } \phi = \sum_{j=1}^{N} \Phi_j b_j, \tag{4.5}$$

with H_j and Φ_j being the value of h and ϕ, respectively, at the grid node with global index $j \in \{1,\ldots,N\}$. Multiplying relation (4.3) with the basis functions as test functions, and integrating over D yields

$$\int_D h b_i \, dx = \int_D \text{div}\left(\frac{\nabla \phi}{\|\nabla \phi\|}\right) b_i \, dx, \quad \forall i = 1,\ldots,N.$$

Next, we plug (4.5) into the above equations, and integrate by parts (see Theorem A.8), obtaining for all $i = 1,\ldots,N$:

$$\sum_{j=1}^{N} H_j \int_D b_j b_i \, dx = -\int_D \frac{\nabla \phi}{\|\nabla \phi\|} \cdot \nabla b_i \, dx$$

$$= -\sum_{j=1}^{N} \int_D \frac{\Phi_j}{\|\sum_{k=1}^{N} \Phi_k \nabla b_k\|} \nabla b_j \cdot \nabla b_i \, dx.$$

Then, the corresponding matrix and right-hand side are assembled elementwise in the usual way (also cf. Chapter 2), and the resulting linear system can be solved for $H = (H_j)_j$.

4.2 Shape Derivative

To solve shape optimization problems such as (1.16) using a steepest descent algorithm, one needs to compute the derivative of its objective functional (1.17) with respect to the shape $\mathcal{O}$. Therefore, a calculus on shapes is needed which is provided by shape sensitivity analysis (see Sokołowski and Zolésio [SZ92], and Delfour and Zolésio [DJPZ01]). Shape sensitivity analysis is a classical subject in shape optimization, and there are two ways of introducing shape sensitivities: the *deformation method* and the *speed* or *velocity method*. For adequately regular shapes $\mathcal{O}$, i.e. those with boundary of class C^1 (cf. [SZ92, DJPZ01]), these two methods are equivalent. In our presentation, we use the speed method and closely follow [BO05]. We will only demonstrate formally how to compute shape derivatives, for a rigorous justification we refer to [SZ92, DJPZ01].

As indicated in Section 4.1, we start with a given shape $\mathscr{O}$ and consider a time evolution of shapes $\mathscr{O}(t)$ according to a velocity field **V**. Then the shapes $\mathscr{O}(t)$ are given as

$$\mathscr{O}(t) = \{x(t) : x(0) \in \mathscr{O}, \dot{x}(\tau) = \mathbf{V}(x(\tau)) \text{ in } (0,t)\}. \tag{4.6}$$

If the velocity field **V** is Lipschitz continuous, i.e. $\mathbf{V} \in C^{0,1}(\mathbb{R}^2; \mathbb{R}^2)$, the Picard-Lindelöf Theorem (see e.g. [Heu06, p. 70]) ensures that $\mathscr{O}(t)$ are well-defined for $t \leq T$. In other words, the shape $\mathscr{O}$ is perturbed by the velocity field **V** according to (4.6).

Suppose we have a shape functional $\mathbf{J}(\mathscr{O})$ such as given by (1.17). Then the shape derivative of **J** in direction of a perturbation $\mathbf{V} \in C^{0,1}(\mathbb{R}^2; \mathbb{R}^2)$ is given by

$$\left. \frac{\mathrm{d}}{\mathrm{d}t} (\mathbf{J}(\mathscr{O}(t))) \right|_{t=0}, \tag{4.7}$$

if the above derivative exists. In that case, we also write $\mathrm{d}\mathbf{J}(\mathscr{O}; \mathbf{V})$ for (4.7), and refer to $\mathrm{d}\mathbf{J}(\mathscr{O}; .)$ as *shape differential*. From the so-called *Hadamard-Zolésio structure theorem* (cf. [DJPZ01, p. 348 ff.]) we know if **J**, $\mathscr{O}$ and the velocity in normal direction $V := \mathbf{V} \cdot n$ are sufficiently smooth, that the shape differential is a linear functional of $V|_{\partial \mathscr{O}}$, such that we also write

$$\mathrm{d}\mathbf{J}(\mathscr{O}; \mathbf{V}) = \langle \mathbf{J}'(\mathscr{O}), V \rangle. \tag{4.8}$$

If $\mathbf{J}'(\mathscr{O})$ is a bounded linear functional on a Hilbert space, we can represent $\mathbf{J}'(\mathscr{O})$ as an element of that space due to the Riesz representation theorem by choosing an appropriate inner product, i.e. a metric, on that space (cf. [Bur03, BO05, dG06]). We will go into further details on what metric we chose in Section 4.4.1.

Since we identified a shape $\mathscr{O}$ with a level set function ϕ via (4.1), we also express the objective functional and the shape sensitivities in terms of the level set function ϕ. More precisely, we introduce the notation

$$\mathscr{J}(\phi) := \mathbf{J}(\{x \in D : \phi(x) < 0\}) = \mathbf{J}(\mathscr{O}).$$

This allows us, taking the above observations and (4.4) into account, to rewrite the shape derivative of **J** in direction **V**, i.e. with velocity $V = \mathbf{V} \cdot n$ in normal direction, in the level set context as follows:

$$\langle \mathbf{J}'(\mathscr{O}), V \rangle = \langle \mathbf{J}'(\{x \in D : \phi(x) < 0\}), -\phi_t \|\nabla \phi\|^{-1} \rangle$$
$$= \langle \mathscr{J}'(\phi), \phi_t \rangle. \tag{4.9}$$

In what follows we describe how shape derivatives of two simple shape functionals can be computed merely formally, following the technique outlined in [BO05, ZCMO96]. With these two prototypes we can derive the shape derivative of our objective functional of interest (1.17) afterwards.

4.2 Shape Derivative

Domain functional. We start with a volume functional of the form

$$\mathbf{J}(\mathcal{O}) := \int_{\mathcal{O}} \varphi(x)\,dx. \qquad (4.10)$$

In the sequel, we write level sets such as $\{x \in D : \phi(x) < 0\}$ concisely as $\{\phi < 0\}$. Then we are looking for an expression for $\langle \mathcal{J}'(\phi), \phi_t \rangle$ which is given with the above observations in mind as

$$\langle \mathcal{J}'(\phi), \phi_t \rangle = \left\langle \mathbf{J}'(\{\phi < 0\}), -\phi_t \|\nabla \phi\|^{-1} \right\rangle$$

$$= \frac{d}{dt}\left(\mathbf{J}(\{\phi(t,.) < 0\})\right)\bigg|_{t=0},$$

with $\{\phi(t,.) < 0\} = \mathcal{O}(t)$. Hence we have to compute $\frac{d}{dt}\mathbf{J}(\mathcal{O}(t))$, which can be done using the Heaviside function $\mathcal{H}$ (cf. Section 3.4) as follows:

$$\frac{d}{dt}\mathbf{J}(\mathcal{O}(t)) = \frac{d}{dt}\int_{\mathcal{O}(t)} \varphi(x)\,dx$$

$$= \frac{d}{dt}\int_{\mathbb{R}^2} \varphi(x)\,\mathcal{H}(-\phi(t,x))\,dx$$

$$= -\int_{\mathbb{R}^2} \varphi(x)\,\delta_0(-\phi(t,x))\,\phi_t(t,x)\,dx.$$

In the last expression, δ_0 denotes the Dirac delta distribution located at 0. The last step is to (formally) apply the coarea formula (see Theorem A.9) to the last expression to obtain

$$\langle \mathcal{J}'(\phi), \phi_t \rangle = \frac{d}{dt}(\mathbf{J}(\mathcal{O}(t)))\bigg|_{t=0}$$

$$= -\int_{-\infty}^{\infty} \left(\int_{\{\phi(0,.)=r\}} \varphi \phi_t \delta_0(-\phi)\|\nabla\phi\|^{-1}\,ds\right) dr$$

$$= -\int_{-\infty}^{\infty} \delta_0(-r)\left(\int_{\{\phi(0,.)=r\}} \varphi \phi_t \|\nabla\phi\|^{-1}\,ds\right) dr$$

$$= -\int_{\{\phi(0,.)=0\}} \varphi \phi_t \|\nabla\phi\|^{-1}\,ds. \qquad (4.11)$$

Note that formula (4.11) coincides with the formula given in [DJPZ01, SZ92] if $-\phi_t \|\nabla\phi\|^{-1}$ is replaced by the velocity in normal direction V according to (4.4).

Boundary functional. The second prototype functional we would like to compute the shape derivative of, is a boundary functional of the form

$$\mathbf{J}(\mathcal{O}) := \int_{\partial \mathcal{O}} \varphi(x)\,ds. \qquad (4.12)$$

To compute the shape derivative of (4.12), we make use of the following identity which allows us to rewrite the boundary functional as a domain functional, such that we can then apply formula (4.11):

$$1 = \frac{\nabla \phi}{\|\nabla \phi\|} \cdot \frac{\nabla \phi}{\|\nabla \phi\|} = \frac{\nabla \phi}{\|\nabla \phi\|} \cdot n. \tag{4.13}$$

Using (4.13) and the divergence theorem gives the following representation for $\mathbf{J}(\mathcal{O})$:

$$\begin{aligned}
\mathbf{J}(\mathcal{O}) &= \int_{\partial \mathcal{O}} \varphi \, ds \\
&= \int_{\partial \mathcal{O}} \varphi \frac{\nabla \phi}{\|\nabla \phi\|} \cdot n \, ds \\
&= \int_{\mathcal{O}} \operatorname{div}\left(\varphi \frac{\nabla \phi}{\|\nabla \phi\|}\right) dx.
\end{aligned} \tag{4.14}$$

Now we simply apply formula (4.11) to (4.14) and obtain, keeping (4.2) and (4.3) in mind,

$$\begin{aligned}
\langle \mathscr{J}'(\phi), \phi_t \rangle &= -\int_{\{\phi=0\}} \operatorname{div}\left(\varphi \frac{\nabla \phi}{\|\nabla \phi\|}\right) \phi_t \|\nabla \phi\|^{-1} ds \\
&= -\int_{\{\phi=0\}} \sum_{i=1}^{2}\left(\varphi_{,i} \frac{\phi_{,i}}{\|\nabla \phi\|} + \varphi \left(\frac{\phi_{,i}}{\|\nabla \phi\|}\right)_{,i}\right) \phi_t \|\nabla \phi\|^{-1} ds \\
&= -\int_{\{\phi=0\}} \left(\nabla \varphi \cdot \frac{\nabla \phi}{\|\nabla \phi\|} + \varphi \operatorname{div}\left(\frac{\nabla \phi}{\|\nabla \phi\|}\right)\right) \phi_t \|\nabla \phi\|^{-1} ds \\
&= -\int_{\{\phi=0\}} (\nabla \phi \cdot n + \varphi h) \phi_t \|\nabla \phi\|^{-1} ds \\
&= -\int_{\{\phi=0\}} \left(\frac{\partial \varphi}{\partial n} + \varphi h\right) \phi_t \|\nabla \phi\|^{-1} ds.
\end{aligned} \tag{4.15}$$

Note again that formula (4.15) coincides with the one derived in [DJPZ01, SZ92] for boundary functionals if $-\phi_t \|\nabla \phi\|^{-1}$ is replaced by the velocity in normal direction V according to (4.4).

We continue with the general objective functional (1.17) given in Definition 1.10. Recall the boundary configuration (1.6) and its properties, and the actual objective functional $\mathbf{J}(\mathcal{O})$ given by

$$\mathbf{J}(\mathcal{O}) = J(\mathcal{O}, u(\mathcal{O})) = \underbrace{\int_{\mathcal{O}} j(u) \, dx + \int_{\partial \mathcal{O}} k(u) \, ds}_{=: \bar{J}_1(\mathcal{O})} + \underbrace{\alpha \int_{\mathcal{O}} 1 \, dx}_{=: \bar{J}_2(\mathcal{O})} + \underbrace{\beta \int_{\partial \mathcal{O}} 1 \, ds}_{=: \bar{J}_3(\mathcal{O})},$$

4.2 Shape Derivative

which we separate into three parts as indicated above, and focus on each of them individually. As introduced in Section 1.1, the only part of the boundary $\partial\mathcal{O}$ to be optimized is Γ_0. Therefore, the perturbations of ϕ at Γ_D and Γ_N are considered 0. Consequently, the integrals appearing in (4.11) and (4.15) reduce to integrals over Γ_0 instead of the whole boundary $\partial\mathcal{O}$.

The shape derivatives of the functionals $\bar{\mathbf{J}}_2$ and $\bar{\mathbf{J}}_3$ in the level set context can be easily obtained by means of (4.11) and (4.15), respectively:

$$\langle \bar{\mathscr{J}}_2'(\phi), \phi_t \rangle = -\alpha \int_{\Gamma_0} \phi_t \|\nabla\phi\|^{-1} \, ds, \tag{4.16}$$

$$\langle \bar{\mathscr{J}}_3'(\phi), \phi_t \rangle = -\beta \int_{\Gamma_0} h\phi_t \|\nabla\phi\|^{-1} \, ds. \tag{4.17}$$

The situation for functional $\bar{\mathbf{J}}_1$ is slightly different. It additionally depends on u which itself depends on the shape $\mathcal{O}$, meaning that we also need to differentiate u with respect to $\mathcal{O}$ — in terms of optimal control problems, we have to differentiate the state with respect to the control. According to Tröltzsch [Trö05, Section 2.10], this can be (formally) avoided by taking the Lagrangian functional into account. We have already introduced the Lagrangian functional in Section 3.1.2 (cf. (3.14)) for the stochastic case. Leaving the stochasticity aside for the time being, the construction of the Lagrangian in Section 3.1.2 can be applied verbatim to $\bar{\mathbf{J}}_1$. In particular, the Lagrangian reads as

$$L(\mathcal{O}, \varphi, \psi) = \int_{\mathcal{O}} j(\varphi) \, dx + \int_{\partial \mathcal{O}} k(\varphi) \, ds + \int_{\mathcal{O}} Ae(\varphi) : e(\psi) \, dx \\ - \int_{\mathcal{O}} f \cdot \psi \, dx - \int_{\Gamma_N} g \cdot \psi \, ds.$$

Now, if u is the solution to the elasticity PDE (1.7) and p the corresponding adjoint state, i.e. solution to (3.20), we know from Section 3.1.2 that the point (u, p) satisfies

$$\bar{\mathbf{J}}_1(\mathcal{O}) = L(\mathcal{O}, u, p)$$

because of Lemma 3.7. Then, according to [Trö05, p. 70], the derivative of $\bar{\mathbf{J}}_1(\mathcal{O})$ with respect to $\mathcal{O}$ can be (formally) obtained by differentiating the Lagrangian $L(\mathcal{O}, u, p)$ with respect to $\mathcal{O}$, i.e. $\langle \bar{\mathbf{J}}_1'(\mathcal{O}), V \rangle = \langle \frac{\partial L}{\partial \mathcal{O}} L(\mathcal{O}, u, p), V \rangle$.

Putting it all together, and having (4.11) and (4.15) in mind, we obtain

$$\langle \bar{\mathscr{J}}_1'(\phi), \phi_t \rangle = -\int_{\Gamma_0} j(u) \phi_t \|\nabla\phi\|^{-1} \, ds - \int_{\Gamma_0} \left(\frac{\partial k(u)}{\partial n} + k(u) h \right) \phi_t \|\nabla\phi\|^{-1} \, ds$$

$$-\int_{\Gamma_0} Ae(u):e(p)\phi_t \|\nabla\phi\|^{-1}\,ds + \int_{\Gamma_0} f\cdot p\phi_t\|\nabla\phi\|^{-1}\,ds$$

$$= -\int_{\Gamma_0}\left(j(u)+\frac{\partial k(u)}{\partial n}+k(u)h+Ae(u):e(p)-f\cdot p\right)\phi_t\|\nabla\phi\|^{-1}\,ds. \quad (4.18)$$

Finally, (4.18), (4.16), and (4.17) yield

$$\langle \mathscr{J}'(\phi),\phi_t\rangle = -\int_{\Gamma_0}\left(j(u)+\frac{\partial k(u)}{\partial n}+k(u)h+Ae(u):e(p)\right.$$
$$\left. -f\cdot p+\alpha+\beta h\right)\frac{\phi_t}{\|\nabla\phi\|}\,ds. \quad (4.19)$$

Remark 4.2. *With formula* (4.19) *we can easily compute the shape derivatives of the special cases introduced in Definition 1.8, namely the compliance* (1.14) *and the quadratic functional* (1.15).

1. Let $\mathbf{J}_1(\mathscr{O})$ be the compliance as defined in (1.14). Then, $j(u)=f\cdot u$, and by Remark 3.10 we have that $k(u)|_{\Gamma_0}=0$, and $p=-u$. (4.19) with $\alpha:=0$ and $\beta:=0$ leads to

$$\langle \mathscr{J}'_1(\mathscr{O}),\phi_t\rangle = -\int_{\Gamma_0}(2f\cdot u-Ae(u):e(u))\phi_t\|\nabla\phi\|^{-1}\,ds. \quad (4.20)$$

2. Let $\mathbf{J}_2(\mathscr{O})$ be the least square error functional defined in (1.15). Then, $j(u)=\frac{1}{2}(u-u_0)^2$ and $k(u)\equiv 0$. In this case (4.19) with $\alpha:=0$ and $\beta:=0$ results in

$$\langle \mathscr{J}'_2(\mathscr{O}),\phi_t\rangle = -\int_{\Gamma_0}\left(\frac{1}{2}(u-u_0)^2+Ae(u):e(p)-f\cdot p\right)\frac{\phi_t}{\|\nabla\phi\|}\,ds. \quad (4.21)$$

At the end of this section, we collect the formulas for the shape derivatives of the objective functionals of the various stochastic shape optimization models introduced in Chapter 3. The stochastic counterpart of the general objective functional $\mathbf{J}(\mathscr{O})$ (cf. (1.17)) is given by $\mathbf{J}(\mathscr{O};\omega)$ defined in (3.3). Recall that the required deformation $u(\mathscr{O};\omega)$ to evaluate $\mathbf{J}(\mathscr{O};\omega)$ for a scenario ω is obtained by solving elasticity PDEs for the basis forces and subsequent assembling of these basis solutions along Algorithm 3.16. Just as before, we denote the level set representation of $\mathbf{J}(\mathscr{O};\omega)$ by $\mathscr{J}(\phi;\omega)$. The shape derivative of $\mathscr{J}(\mathscr{O};\omega)$ is then given by (4.19), where u and p are the corresponding state and adjoint state, respectively, for scenario ω (cf. Theorem 3.13).

4.2 Shape Derivative

Expectation-based model. The random shape optimization problem based on the expectation value was defined in Definition 3.4 (also see (3.37)). Its objective with the above notations in the level set context reads as

$$\mathscr{J}_{\mathbb{E}}(\phi) := \sum_{\mu=1}^{S} \pi_\mu \mathscr{J}(\phi; \omega_\mu). \tag{4.22}$$

Hence, the shape derivative of $\mathscr{J}_{\mathbb{E}}(\mathcal{O})$ is

$$\langle \mathscr{J}'_{\mathbb{E}}(\phi), \phi_t \rangle = \sum_{\mu=1}^{S} \pi_\mu \langle \mathscr{J}'(\phi; \omega_\mu), \phi_t \rangle, \tag{4.23}$$

where $\langle \mathscr{J}'(\phi; \omega_\mu), \phi_t \rangle$ is obtained as mentioned earlier according to (4.19).

Expected excess model. The stochastic shape optimization problem with the expected excess objective functional was defined in Definition 3.17 on page 70. Two different models were introduced in Section 3.3 to solve it numerically. The first one involves a barrier method whereas in the second one, the max-expression in the objective function is smoothly approximated.

Let us start with the barrier model (3.40). Its objective function, i.e. the barrier function, is given by (3.41). In order to avoid confusion with the time parameter t in the level set function, we use $\mathbf{t} \in \mathbb{R}^S$ for the additional variables introduced in (3.39). Then our objective functional in level set notation looks as follows:

$$\mathscr{B}(\mathbf{t}, \phi; \gamma) := \sum_{\mu=1}^{S} \pi_\mu t_\mu - \gamma \left(\sum_{\mu=1}^{S} \ln\left(-\mathscr{J}(\phi; \omega_\mu) + \eta + t_\mu \right) + \sum_{\mu=1}^{S} \ln(t_\mu) \right). \tag{4.24}$$

Both, the shape derivative as well as the derivative with respect to $\mathbf{t}$ of $\mathscr{B}$ will be needed in order to obtain a descent direction. The derivative of $\mathscr{B}$ with respect to $\mathbf{t}$ in direction $\mathbf{t} \in \mathbb{R}^S$ is given as

$$\langle \mathscr{B}_\mathbf{t}(\mathbf{t}, \phi; \gamma), \mathbf{t} \rangle = \sum_{\mu=1}^{S} \left[\pi_\mu t_\mu - \gamma \left(\frac{t_\mu}{-\mathscr{J}(\phi; \omega_\mu) + \eta + t_\mu} + \frac{t_\mu}{t_\mu} \right) \right]. \tag{4.25}$$

The descent direction we use with respect to $\mathbf{t}$ is $\bar{\mathbf{t}} \in \mathbb{R}^S$ defined by

$$\bar{t}_\mu := -\left[\pi_\mu - \gamma \left(\frac{1}{-\mathscr{J}(\phi; \omega_\mu) + \eta + t_\mu} + \frac{1}{t_\mu} \right) \right], \forall \mu = 1, \ldots, S, \tag{4.26}$$

which makes sense, as we can see immediately from (4.25) that $\langle \mathscr{B}_{\mathbf{t}}(\mathbf{t},\phi;\gamma),\bar{\mathbf{t}}\rangle \leq 0$. The shape derivative, i.e. the derivative of $\mathscr{B}$ with respect to $\mathscr{O}$, is given by

$$\langle \mathscr{B}_{\mathscr{O}}(\mathbf{t},\phi;\gamma),\phi_t \rangle = \gamma \sum_{\mu=1}^{S} \frac{\langle \mathscr{J}'(\phi;\omega_\mu),\phi_t \rangle}{-\mathscr{J}(\phi;\omega_\mu)+\eta+\mathbf{t}_\mu}. \qquad (4.27)$$

The second expected excess model (3.42) was described in Section 3.3.2. Its objective functional in level set notation becomes

$$\mathscr{J}_{\mathrm{EE}_\eta}(\phi) := \sum_{\mu=1}^{S} \frac{\pi_\mu}{2} \left(\sqrt{(\mathscr{J}(\phi;\omega_\mu)-\eta)^2 + \varepsilon} + \mathscr{J}(\phi;\omega_\mu) - \eta \right). \qquad (4.28)$$

Therefore, the shape derivative of $\mathscr{J}_{\mathrm{EE}_\eta}$ reads as

$$\langle \mathscr{J}'_{\mathrm{EE}_\eta}(\phi),\phi_t \rangle = \sum_{\mu=1}^{S} \frac{\pi_\mu}{2} \langle \mathscr{J}'(\phi;\omega_\mu),\phi_t \rangle \left(\frac{\mathscr{J}(\phi;\omega_\mu)-\eta}{\sqrt{(\mathscr{J}(\phi;\omega_\mu)-\eta)^2 + \varepsilon}} + 1 \right). \qquad (4.29)$$

Excess probability model. Finally, we consider the excess probability model introduced in Section 3.4, in particular we derive the shape derivative for the objective functional of the smooth approximation (3.44). Recall its objective functional which looks as follows in the level set context:

$$\mathscr{J}_{\mathrm{EP}_\eta}(\phi) := \sum_{\mu=1}^{S} \pi_\mu \frac{1}{1+e^{-2\kappa(\mathscr{J}(\phi;\omega_\mu)-\eta)}}. \qquad (4.30)$$

The shape derivative of $\mathscr{J}_{\mathrm{EP}_\eta}$ is then given by

$$\langle \mathscr{J}'_{\mathrm{EP}_\eta}(\phi),\phi_t \rangle = \sum_{\mu=1}^{S} 2\kappa\pi_\mu \langle \mathscr{J}'(\phi;\omega_\mu),\phi_t \rangle \frac{e^{-2\kappa(\mathscr{J}(\phi;\omega_\mu)-\eta)}}{\left(1+e^{-2\kappa(\mathscr{J}(\phi;\omega_\mu)-\eta)}\right)^2}. \qquad (4.31)$$

Remark 4.3. *In the discrete setting, we have a discrete level set function Φ. Using composite finite elements described in Chapter 2, we can compute the discrete deformation U and the discrete adjoint state P. The discrete mean curvature H is obtained as stated in Section 4.1.1, and the discrete outer normal N is given by $\frac{\nabla \Phi}{\|\nabla \Phi\|}$. Given a discrete descent direction Ξ (cf. Section 4.4.1), the discrete shape derivative can be computed as follows (cf. (4.19)):*

$$\langle \mathscr{J}'(\Phi), \Xi \rangle = -\int_{\Gamma_0} \left(j(U) + \frac{\partial k(U)}{\partial N} + k(U)H + Ae(U):e(P) \right.$$
$$\left. - f \cdot P + \alpha + \beta H \right) \frac{\Xi}{\|\nabla \Phi\|} \, ds.$$

The numerical integration is done via simplicial quadrature rules (cf. Section 2.1.4).

In the case of a stochastic objective functional $\mathscr{J}$, which can be any of the ones presented in this section, we need the discrete deformations $\bar{U}(\omega_s)$ and the discrete adjoint states $\bar{P}(\omega_2)$ for all $s = 1, \ldots, S$ in the above formula. These are obtained according to Algorithm 3.16 by solving elasticity PDEs only for the basis forces.

4.3 Topological Derivative

As already mentioned before, we will employ a descent algorithm to solve our shape optimization problems (cf. Section 4.4 for details). This implies that we need to start our solution method with an initial guess. In particular this means that we can choose initial shapes with different topologies from where the steepest descent method starts. It turns out that the resulting shape at the end of the descent algorithm strongly depends on this choice of initial topology (see e.g. [AJT04, AdGJT05], Fig. 4.2 on page 99 and also Fig. 5.7 on page 107 in Chapter 5). From [AJT04] we know that the level set method in 2D is in general not capable of creating new holes in the structure during the optimization process. However, it can change the topology by closing holes or merging several holes together. Summarizing, one ends up in different local optima depending on the choice of the initial topology.

A remedy would be a pointwise criterion that tells us whether or not it is advantageous for the actual objective functional to take away material at a point. Such a criterion has been introduced with the so-called "bubble method" or *topological sensitivity* by Schumacher [Sch96] for the case of compliance minimization. The essential idea is to perforate the domain by adding a tiny hole, say a circle with radius ρ, and consider the change of objective values for the perforated domain compared to the original one, leading to an asymptotic expansion of a function depending on ρ. The method was generalized to a class of shape functionals by Sokołowski and Żochowski [SZ99] and applied to 3D elasticity in [SZ01]. In [SZ03], the approach is extended to the case of finitely many circular holes, combining topology variations with boundary variations simultaneously. Using an adjoint method and a truncation technique, Garreau et al. [GGM01] computed the topological sensitivity for general objective functionals and arbitrarily shaped holes.

The topological derivative has been incorporated into the level set method (see e.g. [BHR04]) and also combined with the shape derivative in that context (cf. e.g. [AdGJT05, AA06, HKO07]). We also included the topological derivative in our algorithm which allows us to start the optimization process with a solid structure. However, we still cannot be sure that this procedure leads to a globally optimal solution of the underlying shape optimization problem. In particular, the inclusion of the topological derivative adds a few more parameters that need to be chosen, and different choices might lead to different solutions at the end of the descent algorithm (see Figures 5.11 and 5.21 in Chapter 5). The following definition of the topological gradient can be found for example in [GGM01].

Definition 4.4. *Suppose we are minimizing a functional* $\mathbf{J}(\mathcal{O}) = J(\mathcal{O}, u(\mathcal{O}))$. *Let* $\mathcal{O}_\rho = \mathcal{O} \setminus \overline{(x_0 + \rho \mathfrak{M})}$ *be the perforated domain obtained from* $\mathcal{O}$ *by removing a small part* $\mathfrak{M}_\rho := x_0 + \rho \mathfrak{M}$ *from* $\mathcal{O}$. x_0 *is in* $\mathcal{O}$, *and* $\mathfrak{M} \subset \mathbb{R}^2$ *is a fixed open and bounded subset containing the origin. Then, an asymptotic expansion of the function* $\mathbf{J}$ *can be obtained in the following form:*

$$\mathbf{J}(\mathcal{O}_\rho) = \mathbf{J}(\mathcal{O}) + \mathbf{f}(\rho)\mathfrak{T}(x_0) + o(\mathbf{f}(\rho)), \tag{4.32}$$

where $\mathbf{f}$ *is a smooth function with* $\lim_{\rho \to 0} \mathbf{f}(\rho) = 0$, *and* $\mathbf{f}(\rho) > 0$. $\mathfrak{T}(x_0)$ *is called* topological gradient *at the point* $x_0 \in \mathcal{O}$.

The objective function $\mathbf{J}(\mathcal{O}_\rho)$ is computed with the elastic displacement u_ρ which is the solution to

$$\begin{cases} -\operatorname{div}(Ae(u)) = f & \text{in } \mathcal{O}, \\ u = 0 & \text{on } \Gamma_D, \\ (Ae(u))n = g & \text{on } \Gamma_N, \\ (Ae(u))n = 0 & \text{on } \Gamma_0, \\ (Ae(u))n = 0 & \text{on } \partial \mathfrak{M}_\rho. \end{cases}$$

This is the elasticity PDE (1.7) with homogeneous Neumann boundary conditions imposed on the boundary $\partial \mathfrak{M}_\rho$ of the newly created hole. Of course, in the case of stochastic forces, this works exactly the same way, i.e. we would have (3.1) with the same additional homogeneous Neumann boundary conditions on $\partial \mathfrak{M}_\rho$. But for the ease of presentation, we omit the stochasticity for now.

The topological gradient $\mathfrak{T}(x)$ at a point $x \in \mathcal{O}$ provides information for creating a small hole located at x, and can thus be used like a descent direction in the optimization process. Roughly speaking, a hole is created at $x \in \mathcal{O}$ if $\mathfrak{T}(x) < 0$. However, as $\mathfrak{T}$ is defined only on $\mathcal{O}$ and we consider the case that we have no elastic material in $D \setminus \overline{\mathcal{O}}$, we have to be careful not to take away too much material at once, because it will never be added back again in the optimization procedure. This is different in the approach by Amstutz and Andrä [AA06] for example, where

4.3 Topological Derivative

the void areas are simulated by a very soft elastic material, which also allows for the opposite operation, i.e. strengthening of the weak phase.

Before we give the actual formulas for $\mathfrak{T}(x)$ for our objective functionals of interest, we describe how we do a descent step given $\mathfrak{T}(x)$ for all $x \in \mathscr{O}$ with step size control. This procedure will be incorporated into the overall shape optimization algorithm, which is described in Section 4.4, in the following way: After a prespecified number of iterations using the shape derivative alone, a descent step based on the topological derivative is performed, according to the following algorithm (also cf. [AA06]).

Algorithm 4.5 (Descent step using the topological derivative). *Let $\mathscr{O}_k$ be the shape in the current iteration k. Let further s be in $[0,1]$, e.g. $s = 0.9$, and $d := \frac{1-s}{5}$[5]. Moreover, set $c_k := s \min_{x \in \mathscr{O}_k} \mathfrak{T}_k(x)$, where $\mathfrak{T}_k$ denotes the topological gradient computed in the domain $\mathscr{O}_k$. We assume in the following that $c_k < 0$[6]. c_k is a threshold and plays the role of a step size.*

1. *Let $\mathscr{O}_{k+1} := \{x \in \mathscr{O}_k : \mathfrak{T}_k(x) > c_k\}$ be the candidate shape for the next iteration*[7]. *This is achieved numerically when the shapes are represented by level set functions as follows: All values $\phi(x)$ of the level set function at grid points x with $\phi(x) < 0$ and $\mathfrak{T}_k(x) \leq c_k$ are multiplied by -1. Afterwards, ϕ is reinitialized as a signed distance function again, resulting in a level function describing $\mathscr{O}_{k+1}$.*

2. *If $\mathbf{J}(\mathscr{O}_{k+1}) < \mathbf{J}(\mathscr{O}_k)$, accept $\mathscr{O}_{k+1}$ as the next iterate, this algorithm ends here in that case. Otherwise, go to step 3.*

3. *If $s < 1$, set $s \leftarrow s+d$, and update c_k accordingly*[8]. *Go to step 1. Otherwise, this algorithm ends here with $\mathscr{O}_{k+1} := \mathscr{O}_k$*[9].

In what follows, we give the formulas for the topological derivative for the compliance (1.14) and the quadratic functional (1.15). We start, however, with a simple case where the functional **J** does not depend on the state u, namely the volume functional.

---

[5] The denominator 5 in the definition of d is arbitrarily chosen. Other values are possible, essentially this controls the increase of the step size. In our implementation we used the value 5.
[6] Otherwise, there is no descent based on $\mathfrak{T}_k$ possible in this iteration.
[7] All points $x \in \mathscr{O}_k$ with $\mathfrak{T}(x) \leq c_k$ are removed.
[8] Note that d is not updated. This means that step 3 is performed at most 5 times.
[9] In this case, the topological sensitivity information did not lead to any improvement for the objective function, therefore the current shape $\mathscr{O}_k$ is not changed.

Lemma 4.6. *Let* $\mathbf{J}(\mathcal{O}) := \int_{\mathcal{O}} 1 \, dx$. *Then the topological derivative of* $\mathbf{J}$ *is*

$$\mathfrak{T}(x) = -|\mathfrak{M}|. \tag{4.33}$$

Proof. We have for $x_0 \in \mathcal{O}$

$$\begin{aligned}
\mathbf{J}(\mathcal{O}_\rho) &= \int_{\mathcal{O}_\rho} 1 \, dx = \int_{\mathcal{O} \setminus \overline{x_0 + \rho \mathfrak{M}}} 1 \, dx \\
&= \int_{\mathcal{O}} 1 \, dx - \int_{x_0 + \rho \mathfrak{M}} 1 \, dx \\
&= \int_{\mathcal{O}} 1 \, dx - \rho^2 \int_{\mathfrak{M}} 1 \, dx \\
&= \mathbf{J}(\mathcal{O}) - \rho^2 |\mathfrak{M}|,
\end{aligned}$$

and the result follows from the definition (4.32). □

From now on, we take $\mathfrak{M}$ to be the unit ball which simplifies the computations and representations of the topological derivative. Then, the topological derivative for the compliance $\mathbf{J}_1(\mathcal{O})$ and the quadratic functional $\mathbf{J}_2(\mathcal{O})$, both defined in Definition 1.8 on page 15 are given in the following theorem. Proofs for these formulas can be found in [SZ01, GGM01].

Theorem 4.7 (See Theorems 4.3 and 4.4 in Allaire et al. [AdGJT05]). *Let* $\mathfrak{M}$ *be the unit ball of* $\mathbb{R}^2$. *We assume for simplicity that* $f = 0$, *and that* g *and the solution* u *to* (1.7) *are smooth, say* $g, u \in H^2(\mathcal{O}; \mathbb{R}^d)$. *Then, for any* $x \in \mathcal{O}$, *the topological derivative of* $\mathbf{J}_1$ *is*

$$\mathfrak{T}_1(x) = \frac{\pi(\lambda + 2\mu)}{2\mu(\lambda + \mu)} \{4\mu Ae(u) : e(u) + (\lambda - \mu) \operatorname{tr}(Ae(u)) \operatorname{tr}(e(u))\}(x), \tag{4.34}$$

and the topological derivative of $\mathbf{J}_2$ *is given by*

$$\begin{aligned}
\mathfrak{T}_2(x) = &-\frac{\pi}{2}(u(x) - u_0(x))^2 \\
&-\frac{\pi(\lambda + 2\mu)}{4\mu(\lambda + \mu)} \{4\mu Ae(u) : e(p) + (\lambda - \mu) \operatorname{tr}(Ae(u)) \operatorname{tr}(e(p))\}(x),
\end{aligned} \tag{4.35}$$

where p *is the corresponding adjoint state, which is also assumed to be smooth, i.e.* $p \in H^2(\mathcal{O}; \mathbb{R}^2)$, *and defined as the solution to* (3.20).

The observation in the following Lemma tells us that $\mathfrak{T}_1(x) \geq 0$ for all $x \in \mathcal{O}$. As a consequence, this means that Algorithm 4.5 can only result in nucleating holes if there is a volume constraint, i.e. $\alpha > 0$ in (1.17), since then the topological

4.3 Topological Derivative

derivative would read $\mathfrak{T}_1(x) - \alpha\pi$ due to Lemma 4.6 and the fact that $\mathfrak{M}$ is the unit ball in $\mathbb{R}^2$. Note that the topological derivative for the perimeter integral $\int_{\partial\mathcal{O}} 1\,ds$ does not exist (see e.g. [HKO07]), which is why we set $\beta := 0$ in (1.17) in all our computations involving the topological derivative.

Lemma 4.8. $\mathfrak{T}_1(x)$ in (4.34) is always nonnegative.

Proof. The factor $\frac{\pi(\lambda+2\mu)}{2\mu(\lambda+\mu)}$ is always greater than 0 according to Definition 1.1. That is why we focus on the second factor. From (1.5) we know that

$$Ae(u) = 2\mu e(u) + \lambda \operatorname{tr}(e(u))\operatorname{Id}$$
$$= 2\mu e(u) + \lambda \operatorname{div}(u)\operatorname{Id},$$

leading to

$$Ae(u) : e(u) = 2\mu e(u) : e(u) + \lambda \operatorname{div}(u)\operatorname{Id} : e(u)$$
$$= 2\mu e(u) : e(u) + \lambda (\operatorname{div}(u))^2,$$

and

$$\operatorname{tr}(Ae(u)) = \sum_{i=1}^{2}(2\mu e_{ii}(u) + \lambda \operatorname{div}(u))$$
$$= 2(\lambda + \mu)\operatorname{div}(u).$$

Hence, $\operatorname{tr}(Ae(u)) : \operatorname{tr}(e(u)) = 2(\lambda + \mu)(\operatorname{div}(u))^2$. Then, the second factor in (4.34) can be estimated in the following way:

$$4\mu Ae(u) : e(u) + (\lambda - \mu)\operatorname{tr}(Ae(u))\operatorname{tr}(e(u))$$
$$= 8\mu^2 \sum_{i,j=1}^{2} e_{ij}^2(u) + 4\mu\lambda (\operatorname{div}(u))^2 + 2(\lambda - \mu)(\lambda + \mu)(\operatorname{div}(u))^2$$
$$= 8\mu^2 \sum_{i,j=1}^{2} e_{ij}^2(u) + (\operatorname{div}(u))^2 (4\mu\lambda + 2\lambda^2 - 2\mu^2)$$
$$\geq 2\mu^2 \left(4\sum_{i,j=1}^{2} e_{ij}^2(u) - (\operatorname{div}(u))^2\right).$$

A straightforward computation yields

$$2\mu^2 \left(3\left(e_{11}^2(u) + e_{22}^2(u)\right) + 8e_{12}^2(u) - 2e_{11}(u)e_{22}(u)\right)$$

for the last expression above. Next, since $e_{11}(u)e_{22}(u) \leq \frac{1}{2}\left(e_{11}^2(u) + e_{22}^2(u)\right)$, we know that the second factor in (4.34) is greater than or equal to

$$2\mu^2 \left(2\left(e_{11}^2(u) + e_{22}^2(u)\right) + 8e_{12}^2(u)\right) \geq 0,$$

which completes the proof. □

Remark 4.9. *The topological derivatives of our stochastic objective functionals are obtained by replacing the shape derivatives in (4.23), (4.27), (4.29), and (4.31) by the corresponding topological derivatives.*

At the end of this section, we show in Fig. 4.2 how different choices of the parameters α and s (from Algorithm 4.5) affect the influence of the topological derivative, i.e. the amount of material that is cut off.

4.4 Steepest Descent Algorithm

In this section we describe the actual descent algorithm, and the choice and computation of a descent direction in Section 4.4.1. As stated in the beginning of Section 4.2, we assume that the objective **J**, the shapes $\mathcal{O}$, and the velocity in normal direction V of the boundary variations are sufficiently smooth, such that the shape derivative is a continuous linear functional of $V|_{\partial\mathcal{O}}$ on a Hilbert space (cf. (4.8)). Recall the corresponding representation (4.9) of the shape derivative using the level set function, i.e. $\langle \mathcal{J}'(\phi), \phi_t \rangle$. Here the shape functional $\mathcal{J}(\phi)$ can be any of those defined in Section 4.2, i.e. $\mathcal{J}_{\mathrm{E}}(\phi)$ (4.22), $\mathcal{B}(\mathbf{t},\phi;\gamma)$ (4.24), $\mathcal{J}_{\mathrm{EE}_\eta}(\phi)$ (4.28), or $\mathcal{J}_{\mathrm{EP}_\eta}(\phi)$ (4.30). The corresponding shape derivatives can be found in Section 4.2.

In order to minimize the shape functional $\mathcal{J}$, we consider a gradient descent

$$\dot{\phi}(t) = -\operatorname{grad}_{\mathcal{G}} \mathcal{J}(\phi)$$

with respect to a metric $\mathcal{G}$, i.e. inner product, on a suitable Hilbert space $\mathcal{V}$ of variations of the level set function ϕ (cf. [dG06, Pac05]). In Burger and Osher [BO05] and Burger [Bur03], different choices of Hilbert spaces and inner products are proposed and discussed. The gradient $\operatorname{grad}_{\mathcal{G}} \mathcal{J}(\phi) \in \mathcal{V}$ for the inner product $\mathcal{G}$ on $\mathcal{V} \times \mathcal{V}$ is defined in the following way:

$$\mathcal{G}(\operatorname{grad}_{\mathcal{G}} \mathcal{J}(\phi), \xi) = \langle \mathcal{J}'(\phi), \xi \rangle \tag{4.36}$$

for all test functions $\xi \in \mathcal{V}$. Because of the above assumptions, the gradient defined by (4.36) is uniquely determined by the Riesz representation theorem. We

4.4 Steepest Descent Algorithm

specify our metric $\mathscr{G}$ and the computation of the gradient later in Section 4.4.1. For the ease of presentation, we assume that we know how to compute the gradient for now, and continue with the description of the descent algorithm. From (4.36) we know that $-\operatorname{grad}_{\mathscr{G}} \mathscr{J}(\phi)$ is a descent direction for $\mathscr{J}$ since

$$\langle \mathscr{J}'(\phi), -\operatorname{grad}_{\mathscr{G}} \mathscr{J}(\phi) \rangle = -\mathscr{G}(\operatorname{grad}_{\mathscr{G}} \mathscr{J}(\phi), \operatorname{grad}_{\mathscr{G}} \mathscr{J}(\phi)) \leq 0,$$

because $\mathscr{G}$ is an inner product.

In the algorithm we start with an initial level set function ϕ^0, and consider Armijo rule (cf. e.g. [BGLS03]) as a step size control for the time discretization. We start with an initial time step $t^0 > 0$. Then we iteratively compute a sequence of level set functions $(\phi^k)_{k=1,\ldots}$ as follows: In iteration $k \geq 1$, the candidate for the next level set function is given as

$$\phi^k := \phi^{k-1} + t^0 \left(-\operatorname{grad}_{\mathscr{G}} \mathscr{J} \left(\phi^{k-1} \right) \right). \tag{4.37}$$

Next, we test is ϕ^k is accepted by the Armijo rule. This is the case if for a given constant $q \in (0,1)$ the condition

$$\mathscr{J}\left(\phi^k\right) \leq \mathscr{J}\left(\phi^{k-1}\right) + qt^0 \langle \mathscr{J}'\left(\phi^{k-1}\right), -\operatorname{grad}_{\mathscr{G}} \mathscr{J}\left(\phi^{k-1}\right) \rangle \tag{4.38}$$

is satisfied, i.e. if the objective functional decreased sufficiently. This can also be expressed in terms of the metric $\mathscr{G}$ using (4.36):

$$\langle \mathscr{J}'\left(\phi^{k-1}\right), -\operatorname{grad}_{\mathscr{G}} \mathscr{J}\left(\phi^{k-1}\right) \rangle = -\mathscr{G}\left(\operatorname{grad}_{\mathscr{G}} \mathscr{J}\left(\phi^{k-1}\right), \operatorname{grad}_{\mathscr{G}} \mathscr{J}\left(\phi^{k-1}\right)\right)$$
$$= -\frac{1}{(t^0)^2} \mathscr{G}\left(\phi^k - \phi^{k-1}, \phi^k - \phi^{k-1}\right),$$

where the last equality holds because of (4.37). Then (4.38) becomes

$$\mathscr{J}\left(\phi^k\right) - \mathscr{J}\left(\phi^{k-1}\right) \leq -q\frac{1}{t^0}\mathscr{G}\left(\phi^k - \phi^{k-1}, \phi^k - \phi^{k-1}\right). \tag{4.39}$$

If ϕ^k satisfies (4.39), we accept it as the next iterate. Otherwise, we decrease the step size t^0 by multiplying it by some constant $p \in (0,1)$, obtaining a step size t^1. Then we set the candidate $\phi^k := \phi^{k-1} + t^1 \left(-\operatorname{grad}_{\mathscr{G}} \mathscr{J}\left(\phi^{k-1}\right)\right)$, and the repeat the above process.

We summarize the whole procedure, also incorporating the topological derivative, in the following algorithm. We describe it using the discrete counterparts of the occurring continuous ingredients, indicated by using the corresponding capital letters (cf. Notation 2.4).

Algorithm 4.10 (Complete descent algorithm). *The following parameters and ingredients have to be provided:*

- *an initial guess given as a discrete level set function Φ^0;*

- *parameters $p, q \in (0, 1)$ and an initial time step $t^0 > 0$ for the Armijo step size control;*

- *an integer n_{top} that controls how often a descent step based on the topological derivative should be performed (if it is negative, this is never performed);*

- *a positive integer M specifying the total number of iterations to be made[10].*

Set the current iteration $k := 0$.

1. *Compute the deformations $\bar{U}^k(\omega_s)$, and if required by the objective $\mathscr{J}$ also the adjoint states $\bar{P}^k(\omega_s)$, for all scenarios[11] ω_s, $s = 1, \ldots, S$, by solving PDEs only for the basis forces according to Algorithm 3.16 for the current shape Φ^k.*

2. *Compute the discrete descent direction Ξ^k (see Section 4.4.1 how to obtain it).*

3. *Set $l := 0$ and do the following:*

    a) *Set the candidate level set function for the next iteration as $\Phi^{k+1} := \phi^k + t^l \Xi^k$ (cf. (4.37)).*

    b) *Compute $\bar{U}^{k+1}(\omega_s)$ and if necessary $\bar{P}^{k+1}(\omega_s)$ for all $s = 1, \ldots, S$ as before according to Algorithm 3.16 for the shape described by Φ^{k+1}.*

    c) *Check if the Armijo rule (4.39) is satisfied, i.e. whether*

    $$\mathscr{J}\left(\Phi^{k+1}\right) - \mathscr{J}\left(\Phi^k\right) \leq -q \frac{1}{t^l} \mathscr{G}\left(\Phi^{k+1} - \Phi^k, \Phi^{k+1} - \Phi^k\right).$$

    If this is the case, accept Φ^{k+1} as the new iterate, and go to step 4. Otherwise, update the time step size $t^{l+1} := pt^l$, set $l := l+1$, and go to step 3a.

---

[10] A usual convergence criterion would be to check if the shape derivative $\mathscr{J}'(\phi)$ applied to the descent direction is sufficiently small. However, because of numerical discretization errors, this cannot be expected to happen (cf. [AP06]). That is why we chose as in [AP06] a fixed number of iterations in the algorithm. If it turns out that the choice of M was too small, we can simply restart the algorithm with the last shape as initial guess.

[11] This also covers the deterministic case by simply setting $S := 1$.

4.4 Steepest Descent Algorithm

4. If $k \mod n_{top} = n_{top} - 1$, do a nucleation step according to Algorithm 4.5 with Φ^{k+1}. In any case, update $k := k+1$, and if $k \leq M$ go to step 2, otherwise terminate the algorithm.

We can apply the above algorithm directly to our stochastic problems with the following objective functionals: $\mathscr{J}_E(\phi)$ (4.22), $\mathscr{J}_{EE_\eta}(\phi)$ (4.28), and $\mathscr{J}_{EP_\eta}(\phi)$ (4.30). Merely for the model (3.40) with the objective $\mathscr{B}(\mathbf{t}, \phi; \gamma)$ (4.24) we have to incorporate the barrier method, which results in the following algorithm, which essentially wraps an additional layer around Algorithm 4.10.

Algorithm 4.11 (Solving model (3.40) with a barrier method). *Let an initial value for the barrier parameter γ be given, e.g. $\gamma = 1$. Furthermore, we need a factor $C \in (0,1)$ the barrier parameter γ is multiplied by in order to decrease it. We need to provide a stopping criterion in terms of a lower bound for γ, such as e.g. $\underline{\gamma} := 10^{-6}$.*

Additionally, we need initializations required for Algorithm 4.10, i.e. an initial guess Φ^0, a positive integer M (here, a small number such as $M = 5$ usually suffices), an integer $n_{top} \leq M$, and the parameters for the Armijo rule p, q, t^0.

Then we do the following while $\gamma > \underline{\gamma}$:

1. Run Algorithm 4.10 with the objective functional $\mathscr{B}(\mathbf{t}, \phi; \gamma)$ and the derivative

$$\left\langle \mathscr{B}_\mathbf{t}\left(\mathbf{t}^k, \Phi^k; \gamma\right), \bar{\mathbf{t}}^k \right\rangle + \left\langle \mathscr{B}_\phi\left(\mathbf{t}^k, \Phi^k; \gamma\right), \Xi^k \right\rangle,$$

given by (4.25), (4.26), and (4.27), respectively. Ξ^k denotes the discrete descent direction computed in step 2 of Algorithm 4.10. Step 3a in Algorithm 4.10 needs to be extended by

$$\mathbf{t}^{k+1} := \mathbf{t}^k + t^l \bar{\mathbf{t}}^k,$$

to update the next candidate for $\mathbf{t}$. Note that $\mathbf{t}^0$ can be initialized in step 3b of Algorithm 4.10 as

$$t_i^0 := \mathscr{J}(\Phi^0; \omega_i) - \eta + 0.1,$$

such that it is feasible for (3.39).

2. Update $\gamma := C\gamma$.

4.4.1 Regularized Descent Direction

This section is dedicated to the final missing piece in Algorithm 4.10, namely the computation of the discrete descent direction Ξ. If Φ is the discrete level set function describing the current shape, Ξ is uniquely determined by (cf. (4.36))

$$\mathscr{G}(\Xi,\xi) = -\langle \mathscr{J}'(\Phi),\xi \rangle \tag{4.40}$$

for all test functions ξ. We still need to specify the metric $\mathscr{G}$ we used. The support of the shape derivative is contained in Γ_0 (see (4.19)). Therefore we take a regularized gradient descent into account, in particular using the metric

$$\mathscr{G}(\zeta,\xi) = \int_D \zeta\xi + \frac{\varsigma^2}{2}\nabla\zeta\cdot\nabla\xi\,ds, \tag{4.41}$$

which is related to a Gaussian filter with width ς. With this, the descent direction Ξ, i.e. the update function for the current level set function Φ according to step 3a in Algorithm 4.10, is not only defined on Γ_0 but on the whole working domain D. This metric ensures smoothness of the descent path and is expected to approximate a regular minimizer from the set of all minimizers.

To compute Ξ we test (4.40) with piecewise linear continuous standard finite element functions on D (see the beginning of Chapter 2) that vanish on $\Gamma_D \cup \Gamma_N$. This procedure yields a system of linear equations that is solved quickly using a cg solver.

In other words, we have to solve a linear elliptic problem of the type

$$\left(\mathrm{Id} - \frac{\varsigma^2}{2}\Delta\right)\phi = r$$

to obtain the descent direction in each time step. The right-hand side r consists of $-\langle \mathscr{J}'(\phi),\xi \rangle$ which is computed in the discrete setting as stated in Remark 4.3. ς is chosen depending on the grid discretization, in most of our computations we set $\varsigma := 4h$, where h denotes the grid discretization parameter from Chapter 2. Fig. 4.3 shows the descent directions computed for the set-up depicted in Fig. 2.10(a) and two different choices for ς.

4.4 Steepest Descent Algorithm

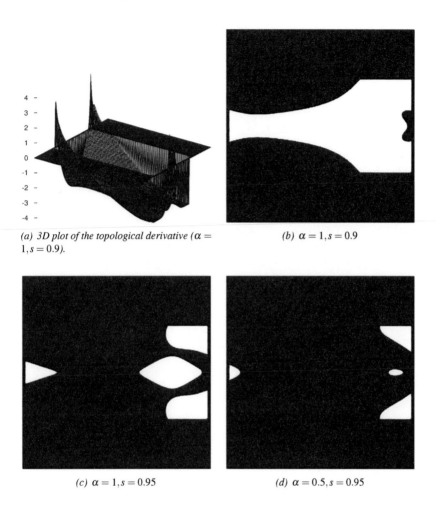

(a) 3D plot of the topological derivative ($\alpha = 1, s = 0.9$).

(b) $\alpha = 1, s = 0.9$

(c) $\alpha = 1, s = 0.95$

(d) $\alpha = 0.5, s = 0.95$

Fig. 4.2: Some effects of different choices for the parameters α and s from Algorithm 4.5 for the topological derivative in the case of compliance minimization are shown. The whole working domain D is shown, using the set-up from Fig. 2.10(a). The white areas show the parts of the domain which would be cut off in step 1 of Algorithm 4.5.

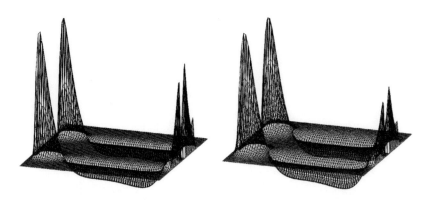

Fig. 4.3: Descent directions in the case of compliance minimization without volume forces obtained for the metric $\mathscr{G}$ (4.41) with $\varsigma = 2h$ on the left, and $\varsigma = 4h$ on the right. The whole working domain D is shown, and the set-up can be found in Fig. 2.10(a).

5 Numerical Results

We finally present various numerical results in this chapter. Our main intention is to demonstrate that the results obtained by our stochastic approach differ significantly from those where, for example, all random variables are simply replaced by their expectations (especially see Section 5.1.1 in this context). Therefore, we particularly consider shape optimization applications where the desired behavior can be observed. In Section 5.1, we present some deterministic problem set-ups, together with some two-stage stochastic optimization counterparts based on the expectation value (cf. Definition 3.4). Subsequently, some first results for shape optimization problems with risk objectives (cf. Sections 3.3 and 3.4) are reported in Section 5.2. All computational results are obtained by Algorithm 4.10 or Algorithm 4.11, respectively, with the appropriate shape objective functional and its derivative as given in Section 4.2.

Recall the general set-up we assumed throughout this work, in particular the boundary partition (1.6): The Dirichlet boundary Γ_D is held fixed during the optimization, as is the Neumann boundary Γ_N, which means that Γ_0 is the only part of $\partial \mathcal{O}$ to be optimized. In all instances, we assume that the surface loads g, or $g(\omega)$ in the stochastic case, act on Γ_N. The actual configurations, i.e. the set of forces constituting the scenarios, are indicated in the figures by arrows, such as for example in Fig. 5.2, where the forces act on the whole top edge of the depicted carrier plate. Sometimes, especially if the number of scenarios is rather big, we show the individual surface loads acting on a half circle instead, as for example in Fig. 5.1, Fig. 5.3, and Fig. 5.4. In these cases, the forces are also understood to be acting on the complete upper edge of the drawn square. The length of the arrows is determined by the force's intensity weighted with the corresponding probability π_σ of that scenario. Note that, unless stated otherwise, we assume that there are no volume forces, i.e. $f \equiv 0$, and that we are minimizing the compliance (1.14) with $\alpha > 0$ and $\beta = 0$ in (1.17) (cf. Remark 1.11).

For purely aesthetical reasons, we keep the level set function fixed on small rectangular boxes right next to Γ_N and Γ_D for all of our instances. In particular, we fixed the level set function on a neighborhood of Γ_N and Γ_D with width 0.03. This is indicated in the configuration sketches, such as in Fig. 5.1 on the left, by the hatched boxes next to Γ_N and Γ_D. Of course, the elastic deformations are still computed in those areas, but the level set function is not changed in a descent step.

For some instances, we color-coded our obtained optimal results according to the von Mises stress. In all these cases, we used a color scale increasing from blue to red, with blue corresponding to the value 0 (cf. Fig. 5.2 (left)). In all energy plots that depict the progression of the volume, it actually shows the volume already scaled by the penalization parameter α.

5.1 Deterministic and Expectation-Based Results

Our first few instances consist of optimizing a carrier plate which is fixed at the bottom and subjected to surface loads on the top (cf. the initial guess in Fig. 5.1). The bottom edge is the Dirichlet boundary Γ_D, whereas the complete upper edge consists of Γ_N, where the surface loads act on. We are looking for an optimal construction between the top edge and the bottom one.

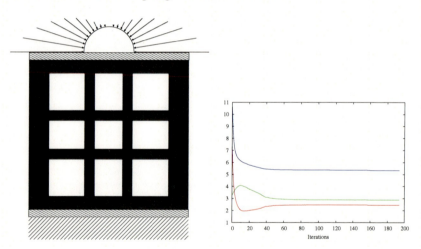

Fig. 5.1: The initial domain used to start Algorithm 4.10 in the computation of the optimal shapes in Fig. 5.2, Fig. 5.3 and Fig. 5.4 is depicted on the left. On the right the different contributions to the objective function are plotted over the number of iterations. The blue curve shows the robust decay of the actual objective functional, whereas the red curve and the green curve display the evolution and the interplay of the compliance functional and the volume term, respectively.

At first, we consider stochastic loadings with two scenarios as shown in Fig. 5.2 on the right. Obviously, we need two basis forces g_1 and g_2, i.e. $K_2 = 2$ in (3.22),

5.1 Deterministic and Expectation-Based Results

to obtain $g(\omega_1)$ and $g(\omega_2)$ from Fig. 5.2 as linear combinations. We assume that the scenarios ω_1 and ω_2 occur equally likely, hence we set $\pi_1 = \pi_2 = \frac{1}{2}$. Since we have already prescribed holes in the initial guess shown in Fig. 5.1, we did not use the topological derivative in this configuration. This can be achieved for example by setting $n_{\text{top}} = -1$ in Algorithm 4.10. The result we obtained at the end of Algorithm 4.10 can be seen in Fig. 5.2 on the right. The symmetric, x-shaped construction should be able to sustain both of the two possible loading configurations equally well. Also note that the level set method was able to change the topology during the optimization process in that the number of holes is decreased from initially 9 to 4 in the end. We directly compared our two-stage approach

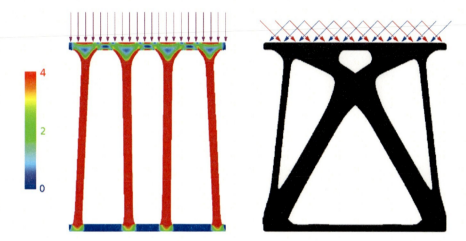

Fig. 5.2: A direct comparison of two-stage stochastic optimization and deterministic optimization for an averaged load is shown. On the right, one can see the optimal shape obtained as the solution of the stochastic model (3.7) with two scenarios ω_1, ω_2, indicated by the differently colored arrows representing surface loads $g(\omega_1)$ and $g(\omega_2)$ acting on the upper plate, both occurring with probability $\frac{1}{2}$. On the left the optimal shape color-coded with the von Mises stress is drawn for the deterministic load $\frac{1}{2}g(\omega_1) + \frac{1}{2}g(\omega_2)$.

with the deterministic case resulting from simply replacing all random variables, i.e. the two surface loads, by their expectation value. The resulting averaged force is pointing straight downwards, as shown in Fig. 5.2 on the left. Consequently, the outcome of the optimization algorithm consists of almost parallel pillars, and as such differs significantly from the stochastically optimal result; also see Section 5.1.1 for a further discussion of this instance.

Next, we consider the same type of boundary configuration and initial shape, but with different stochastic loadings. More precisely, we consider one case with 20 scenarios as shown in Fig.5.3, and one with 21 scenarios as shown in Fig. 5.4. One scenario consists of one of the depicted forces, and each of them acts on the whole upper plate. The resulting optimal shapes can be seen in Fig. 5.3 and Fig. 5.4, respectively, where the von Mises stress distribution is shown for the first ten scenarios. The asymmetric choice of scenarios in Fig. 5.3 yields, as expected, a slightly asymmetric optimal shape, whereas the 21 totally symmetric scenarios from Fig. 5.4 also lead to a totally symmetric result at the end of Algorithm 4.10. Note that two basis forces suffice again to combine all of the stochastic forces linearly out of these. For the set-up from Fig. 5.4, we plotted the objective functional in the course of iterations, split into volume part and compliance part, in Fig. 5.1.

The parameters we used for the computations shown in figures 5.2, 5.3, and 5.4 are the following: As mentioned above, we minimize the compliance plus a weighted volume term. We chose the weight of the volume $\alpha = 8$, and set the Lamé coefficients $\lambda = \mu = 40$, as in all of our presented instances. All these computations were done on a uniform grid of triangles as introduced in Chapter 2 with $(2^8+1) \times (2^8+1)$ nodes. This results in a grid discretization parameter $h = 2^{-8}$ in Chapter 2. The remaining parameters required for Algorithm 4.10 are $n_{\text{top}} = -1$, which means that we are not using the topological derivative at all, and $\varsigma = 6h$ in figures 5.3 and 5.4, and $\varsigma = 4h$ in Fig. 5.2 for the computation of the descent direction (cf. Section 4.4.1). Finally, we used $q = 0.2$, $p = 0.5$, and $t^0 = h$ as parameters for the Armijo rule in Algorithm 4.10.

Since the optimal results from figures 5.3 and 5.4 seemingly do not differ very much, we compared the objective values of the two corresponding objectives for the two obtained optimal shapes. This can be seen in Fig. 5.5, and it clearly shows that the symmetric solution is not as good as the asymmetric one for the non-symmetric configuration, and vice versa.

As a second application, we consider the shape optimization of a cantilever, which can be found quite frequently in the relevant literature. The initial domain is given in Fig. 5.6 on the left, together with a deterministic surface load pointing downwards, which is applied on the small marked part in the middle of the right edge. The structure is fixed on the opposite left side. On the right in Fig. 5.6, the resulting optimal shape is shown, color-coded with the von Mises stress again. We did not make any use of the topological derivative so far. In order to show the dependence of the outcome of Algorithm 4.10 on the choice of the initial domain, we let it run with varying initial shapes as shown in Fig. 5.7. This figure also shows the resulting optimal shapes, and it particularly demonstrates that if we do not prescribe any holes, we will consequently end up with a shape without holes.

5.1 Deterministic and Expectation-Based Results

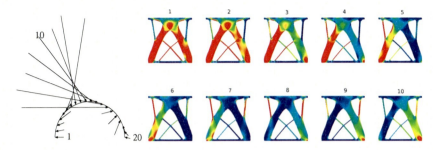

Fig. 5.3: Stochastic shape optimization based on 20 scenarios is depicted. On the left the different loads $g(\omega_\sigma)$ with probabilities π_σ are sketched. Each arrow represents one scenario where the arrow length is determined by the corresponding force intensity weighted with the probability π_σ of the corresponding scenario. On the right the von Mises stress distribution is color-coded on the optimal shape for the first 10 out of the 20 realizations of the stochastic loading. Due to the non-symmetric loading configuration the resulting shape is asymmetric as well. In particular, the right carrier is significantly thicker than the left one, whereas the connecting diagonal stray pointing up right is thinner than the one pointing down left.

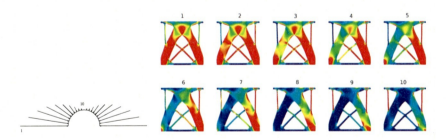

Fig. 5.4: Results for a symmetric load configuration with 21 scenarios, to be contrasted with those reported with an asymmetric configuration in Fig. 5.3. Again, on the left the configuration is sketched, and on the right the von Mises stress distribution is plotted for the first 10 scenarios.

However, the energy plots belonging to the middle and the right column in Fig. 5.7 show that the objective values at the end of the algorithm are very close to each other, although the obtained solutions are not the same. This suggests that there are several local minima found by our algorithm, depending on the initial guess. We also report in Fig. 5.8 how the result depends on varying the volume penalization parameter α.

	$\mathscr{O}_1$	$\mathscr{O}_2$
objective from Fig. 5.3	4.32398	4.4342
objective from Fig. 5.4	5.54182	5.35328

Fig. 5.5: Let $\mathscr{O}_1$ denote the optimal shape from Fig. 5.3, and $\mathscr{O}_2$ the one from Fig. 5.4. The table shows the cost functionals arising from the different stochastic loadings shown in Fig. 5.3 and Fig. 5.4, respectively, evaluated at $\mathscr{O}_1$ and $\mathscr{O}_2$.

We used the same parameter settings as for the carrier plate instances, except for $\varsigma = 4h$ for the regularized descent direction, and $\alpha = 0.3$. With these parameters, we also considered a stochastic cantilever instance with 21 scenarios. The stochastic configuration and the resulting optimal shape can be found in Fig. 5.9. Again, two basis forces are enough to obtain all of the depicted loads.

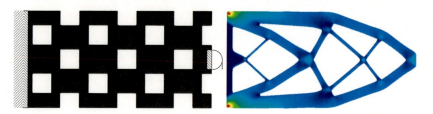

Fig. 5.6: The initial domain for the computation in case of a cantilever geometry is rendered on the left. The left boundary is a Dirichlet boundary where the cantilever is attached to a vertical wall. The center part of the right boundary is the support Γ_N of the a boundary force, which is a deterministic downward-pointing force in this sketch. The resulting optimal shape computed by the proposed level set algorithm is plotted on the right and color-coded with the von Mises stress. A stochastic set-up is reported in Fig. 5.9.

If we use the topological derivative, it is possible to start Algorithm 4.10 with the solid initial shape shown in Fig. 5.7 on the top left. The resulting optimal shape, when setting $n_{\text{top}} = 15$ in the algorithm together with the corresponding energy plots are shown in Fig. 5.10. Here we set $\alpha = 0.4$ and used a grid with $(2^9 + 1) \times (2^9 + 1)$ nodes. For the cut-off threshold in Algorithm 4.5 we used $s = 0.9$. In Fig. 5.11, we ran the same instance again, but this time with $n_{\text{top}} = 5$, giving the shape shown in the top row. The optimal result looks slightly different than the one in Fig. 5.10, and also the objective value is slightly better. However, setting $n_{\text{top}} = 15$ and $s = 0.95$, the algorithm ends with a shape which actually

5.1 Deterministic and Expectation-Based Results

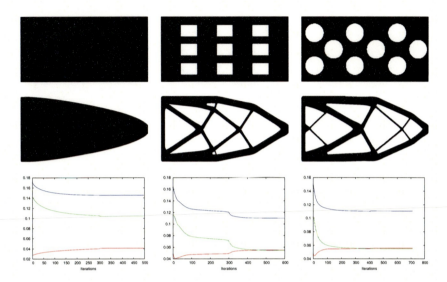

Fig. 5.7: Results for different initial shapes for the deterministic cantilever computation (see Fig. 5.6). The top row shows the initial guess. The corresponding optimal shapes and energy plots are depicted in the second and third row, respectively. In all cases, α is fixed to 0.3. The middle and right simulation results are obviously local minima with values of the cost functional that are fairly close, as indicated by the objective plot.

yields a worse objective value, which can be also seen in Fig. 5.11. This shows already that the parameters for Algorithm 4.5 have to be chosen with care. Moreover, it shows that the inclusion of the topological derivative into the algorithm does not guarantee a global optimal solution.

Next, we consider a set-up where a long rectangle is fixed at its bottom edge, and surface loads may act on two distinct areas on the top edge (cf. Fig. 5.12). In Fig. 5.12 there are 10 scenarios shown, where 5 of these act on the left upper part and the other 5 on the right upper part. All of them occur with equal probability. This configuration requires 4 basis forces: two of them have their support only on the upper left part, whereas the other two's support lies on the upper right side. Figures 5.13 and 5.14 show two different deterministic selections of the 10 scenario forces in Fig. 5.12, and the corresponding optimal results and energy plots. In all three computations we used the parameters $p = 0.5$, $q = 0.1$, $\varsigma = 4h$ on a $(2^8+1) \times (2^8+1)$ grid. Furthermore, we chose $n_{\text{top}} = 5$, $s = 0.9$, and $\alpha = 1$ in Fig. 5.12 and $\alpha = 0.5$ in Fig. 5.13 and Fig. 5.14.

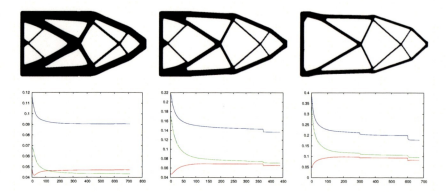

Fig. 5.8: Results for variations of the volume penalization parameter α. In all shown test runs, the initial shape shown in Fig. 5.7 on the right was used. From left to right, the optimal solutions correspond to the choices $\alpha = 0.2, \alpha = 0.5, \alpha = 1$.

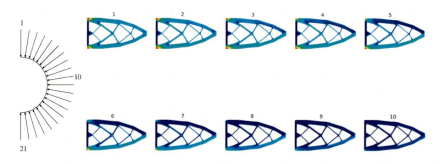

Fig. 5.9: Stochastic shape optimization in the cantilever case with 21 scenarios. The different loads $g(\omega_\sigma)$ with probabilities π_σ are sketched on the left. The von Mises stress distribution is color-coded on the stochastically optimal shape for the first 10 out of the 21 scenarios.

Since we set the boundary regularization parameter $\beta = 0$ (cf. Remark 1.11), the shapes tend to become rather complicated, especially close to corners, during the optimization process. This effect can be seen in Fig. 5.15 and Fig. 5.16 on left, respectively. We propose two ways to cope with this issue. The first one is to apply a morphological operator based on *erosion* and *dilation* known from image processing (cf. e.g. [Soi03] for details), which can be represented by partial differential equations. These can then be solved efficiently using a fast-marching algorithm

5.1 Deterministic and Expectation-Based Results

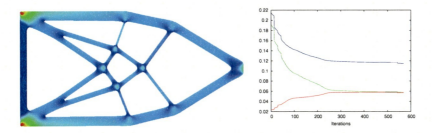

Fig. 5.10: The optimal shape for a cantilever problem with deterministic loading is computed based on the combined shape derivative and topological derivative approach. We chose $n_{\text{top}} = 15$ in Algorithm 4.10. The corresponding energies, i.e. the total value of the objective function (blue), the enclosed volume (green), and the compliance functional (red), are plotted on the right.

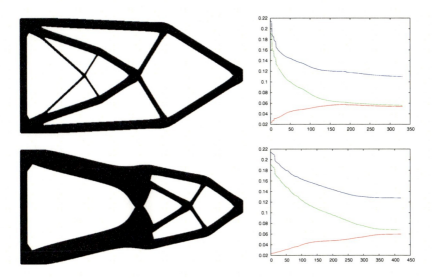

Fig. 5.11: In these instances, we used the same settings and configuration as in the cantilever computation in Fig. 5.10. For the computation in the top row, we set $n_{\text{top}} = 5$, and one can observe that the resulting optimal shape has a slightly better objective value than the one in Fig. 5.10. For the bottom row, we set $n_{\text{top}} = 15$ again, but chose $s = 0.95$ instead of $s = 0.9$ in contrast to Fig. 5.10. The resulting shape looks a lot different, and also its objective value is worse compared to the other parameter settings.

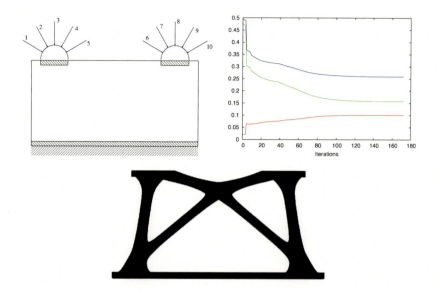

Fig. 5.12: A stochastic set-up with 10 scenarios. The topological derivative was used every $n_{\text{top}} = 5$ iterations. The decay of the objective is drawn in blue, the compliance in red, and the enclosed volume in green. The second row shows the optimal result at the end of Algorithm 4.10.

(cf. [AS99, OS88]). We denote the discrete dilation operator by $\mathbf{D}(.)$, and the discrete erosion operator by $\mathbf{E}(.)$, which take a width parameter as arguments. Then, at the end of Algorithm 4.10 or Algorithm 4.11, we apply the morphological operator $\mathbf{D}(0.5h)\,\mathbf{E}(h)\,\mathbf{D}(0.5h)$, and restart the algorithm with the resulting smoother shape. In Fig. 5.15, one can see the result after such an operation on the right.

The other way is simply to set the parameter β to something greater than 0, and restart Algorithm 4.10 or Algorithm 4.11 with $n_{\text{top}} = -1$, such that the topological derivative is no longer used[1]. After a few iterations, the shape should be much smoother, as demonstrated in Fig. 5.16 on the right. Note, however, that this approach does not work in case of the excess probability objective functional, as such small changes in shape usually do not push the objective value above or below the threshold value η such that the Heaviside function does not change. Also, since

---

[1] Recall from Section 4.3 that the perimeter functional is not topologically differentiable.

5.1 Deterministic and Expectation-Based Results 111

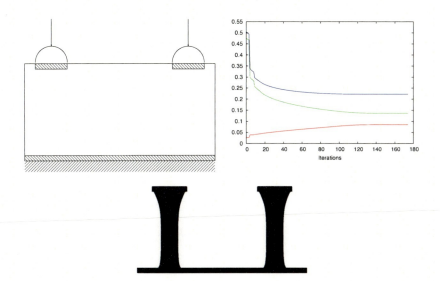

Fig. 5.13: Here we have the same boundary configuration as in Fig. 5.12, however, we consider only the two indicated deterministic forces. The resulting optimal shape is clearly not optimal any more if the straight downward pointing forces are perturbed slightly, and it would therefore perform really badly for the stochastic set-up of Fig. 5.12.

we have to do a few iterations which involve solutions of elasticity PDEs, this approach in general takes longer in computation time compared to the morphological approach described above.

At the end of this section, we give one result obtained by minimizing a quadratic functional instead of the compliance. This instance consequently also requires the computation of the adjoint states. More precisely, we consider the objective functional

$$\int_{\mathcal{O}} Fu^2 \, dx + \alpha \int_{\mathcal{O}} 1 \, dx,$$

where $F \in L^{\infty}(D)$ is a nonnegative weight factor. We consider the same stochastic set-up given in Fig. 5.2 on the right, i.e. two scenarios that are equally likely. We set $\alpha = 0.1$ and F to 1 on the whole hatched box next to the upper plate in Fig. 5.2, and 0 everywhere else. That way, we expect a similar optimal structure as shown in Fig. 5.2 on the right. This is indeed the case, as can be seen in Fig. 5.17. All the other parameters were exactly the same as in the corresponding instance with the compliance as objective functional.

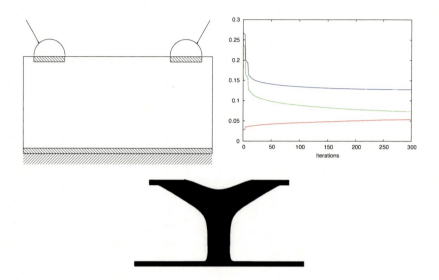

Fig. 5.14: Similar to Fig. 5.13, we have another selection of two of the stochastic forces given in Fig. 5.12, which are considered deterministic here. Again, the shown optimal shape cannot be optimal any more for a slightly perturbed configuration of surface loads.

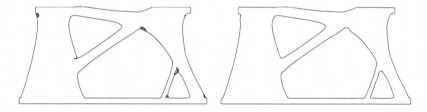

Fig. 5.15: Here we demonstrate the effect of the morphological smoothing operator $\mathbf{D}(0.5h)\,\mathbf{E}(h)\,\mathbf{D}(0.5h)$. It was applied to the shape obtained at the end of Algorithm 4.10, which is shown here on the left, and lead to the much smoother shape shown on the right. Then, we can start Algorithm 4.10 again with the smoother shape as initial guess. The final result can be seen in Fig. 5.22.

5.1.1 VSS and EVPI

Stochastic programs are known to be computationally hard to solve, which raises the question whether the additional effort pays off compared to solving simpler

5.1 Deterministic and Expectation-Based Results

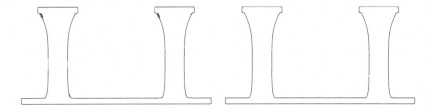

Fig. 5.16: Instead of the morphological regularization based on erosion and dilation (cf. Fig. 5.15), we can also add a perimeter penalization to the objective functional. Precisely, Algorithm 4.10 ended with the result shown on the left, and we restarted the algorithm with $n_{\text{top}} = -1$, i.e. no descent step based on the topological derivative is done, and $\beta = 0.1$ (recall the general objective functional (1.17)). After 16 iterations, we ended up with the much smoother shape shown on the right. This is the final result from Fig. 5.13.

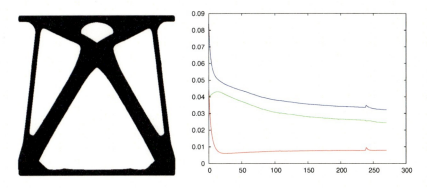

Fig. 5.17: The same stochastic set-up as in Fig. 5.2 was considered in this instance. However, the objective was not the compliance this time but a quadratic functional. The result looks very similar to the one from Fig. 5.2 on the right, as expected. In the energy plot, the red line shows the progression of the quadratic functional, whereas the blue one is still the total objective functional, and the green line shows the progression of the enclosed volume (already scaled by $\alpha = 0.1$). Note that the bump in the graph of the objective function at iteration 240 is due to a morphological smoothing step (cf. Fig. 5.15).

deterministic problems. In particular, there are two common concepts to measure the quality of the stochastic solution, namely the *Value of the Stochastic Solution* (VSS) and the *Expected Value of Perfect Information* (EVPI) (cf. [BL97] for details). We computed these two values for the instance shown in Fig. 5.2.

The optimal objective value of the *recourse problem* (3.7) is denoted by RP, and we consider the following deterministic program, which is called *expected value problem*:
$$\mathrm{EV} := \min\{\mathbf{J}(\mathscr{O};\bar{\omega}) : \mathscr{O} \in \mathscr{U}_{\mathrm{ad}}\},$$
where $\bar{\omega}$ indicates that all occurring random variables are substituted by their expectations. Let $\mathscr{O}_{\mathrm{EV}} \in \arg\min\{\mathbf{J}(\mathscr{O};\bar{\omega}) : \mathscr{O} \in \mathscr{U}_{\mathrm{ad}}\}$. Note that in our example, $\mathscr{O}_{\mathrm{EV}}$ is shown in Fig. 5.2 on the left. Next, we can define the *expected result of using the EV solution* as $\mathrm{EEV} := \sum_{i=1}^{S} \pi_i \mathbf{J}(\mathscr{O}_{\mathrm{EV}};\omega_i)$, which finally leads to the VSS given by $\mathrm{VSS} = \mathrm{EEV} - \mathrm{RP}$. For our particular instance, we have $\mathrm{VSS} = 53.68$, or about 94 % of the EEV.

To compute the EVPI, we have to compute the so-called *wait-and-see* solution WS. If $\mathscr{O}_i$ for $i = 1,\ldots,S$ denote the solutions to the problems
$$\min\{\mathbf{J}(\mathscr{O};\omega_i) : \mathscr{O} \in \mathscr{U}_{\mathrm{ad}}\}, \quad i = 1,\ldots,S,$$
which amounts to solving as many problems as there are scenarios, then WS is defined to be $\mathrm{WS} := \sum_{i=1}^{S} \pi_i \mathbf{J}(\mathscr{O}_i;\omega_i)$, and $\mathrm{EVPI} := \mathrm{RP} - \mathrm{WS}$. For our instance, we obtained $\mathrm{EVPI} = 0.24$.

It is intuitively not surprising that the VSS is so big in our case. The optimal shape shown on the left in Fig. 5.2, $\mathscr{O}_{\mathrm{EV}}$ in the above notation, is clearly far from optimal if any other than the vertical force is applied to it, such as the two diagonal forces depicted on the right in Fig. 5.2 constituting the two scenarios. This is why it clearly pays off to solve the stochastic model in this case.

Fig. 5.2 shows the obtained solution to the recourse problem RP on the right. Roughly speaking, it consists of almost diagonal bars crossing each other. Such a diagonal structure can also be expected if only one of the two forces occurs, in particular in direction of that one force. This would correspond to the wait-and-see solution, and consequently WS and RP should not be overly different — which can be observed in the small value obtained for the EVPI.

5.2 Risk Aversion

This last section is dedicated to some first results in the case of expected excess and excess probability as objectives. All the instances in this section share the following set of parameters:

- a grid with $(2^8 + 1) \times (2^8 + 1)$ nodes,
- $p = 0.5, q = 0.1$, and $t^0 = h$ for the Armijo rule,

5.2 Risk Aversion

- if applicable, an initial barrier parameter $\gamma = 1$,
- $n_{\text{top}} = 5$ (except for Figures 5.18 and 5.19, where $n_{\text{top}} = -1$),
- $\varepsilon = 0.1$ in the expected excess objective (3.42),
- $\kappa = 10$ in the excess probability objective (3.44).

At first, we consider an instance with no surface loads. The tower-like initial shape and the set-up of basis volume forces can be found in Fig. 5.18. We assume that we have three scenarios, constituted by the coefficients (recall (3.22)) $\left(c_1^f(\omega_1), c_2^f(\omega_2), c_3^f(\omega_3) \right)$ given as follows, together with the corresponding probabilities:

$$\omega_1 : (1,0,0) \qquad \pi_1 = 0.45$$
$$\omega_2 : (1,1,0) \qquad \pi_2 = 0.45$$
$$\omega_3 : (1,0,2) \qquad \pi_3 = 0.1.$$

The optimal results with $\alpha = 0.4$ and $\eta = 0.1$ can be found in Fig. 5.19. Compared to the expectation-based result, the shapes obtained solving the approximated expected excess model, as well as the one obtained from solving the barrier problem, are slightly wider, especially on the left side. One can see certain similarities in the two results for the expected excess problem, however, the shapes of the holes differ significantly.

The next instance uses essentially the same configuration as the one from Fig. 5.12. We consider surface loads again, and the first five constituting the first five scenarios are less likely, i.e. $\pi_i = 0.01, i = 1, \ldots, 5$, than the last five ($\pi_i = 0.19, i = 6, \ldots, 10$). The results obtained from Algorithm 4.10 solving problem (3.42) are shown in Fig. 5.20. We chose $\alpha = 1$, and $\eta = 0.4$ for the top computation in Fig. 5.20, and $\eta = 0.6$ for the one in the middle. The cut-off threshold for the topological derivative was $s = 0.9$ in Algorithm 4.5. Compared to the solution to the expectation-based model, which is also shown in Fig. 5.20 at the bottom, the shapes obtained as the solutions to the expected excess model should be able to sustain the unlikely but possibly present forces $g(\omega_1), g(\omega_2), g(\omega_3), g(\omega_4), g(\omega_5)$ significantly better.

We also solved this instance using the barrier approach and Algorithm 4.11 with $\alpha = 1$, $n_{\text{top}} = 5$, and $\eta = 0.4$. The results can be found in Fig. 5.21, on the left with $s = 0.8$ and on the right with $s = 0.9$ in Algorithm 4.5. The results look quite different compared to the one given in Fig. 5.20. The objective value of model (3.39) in case of $s = 0.8$ is 0.00395, and in case of $s = 0.9$ about 0.00808.

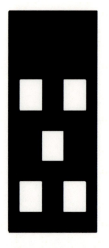
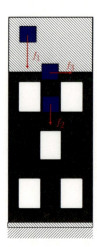

Fig. 5.18: The initial shape (left) and the configuration of basis volume forces (right) used in the first instance with the expected excess objective. This time there are no surface loads. The three basis volume forces f_1, f_2, and f_3 are indicated by the red arrows. f_1 acts on the upper part of the tower (on the whole hatched area), f_2 acts on the lower part, and finally f_3 acts on the complete body. The Dirichlet boundary Γ_D is located at the bottom edge, and the hatched areas are kept fixed during the optimization process.

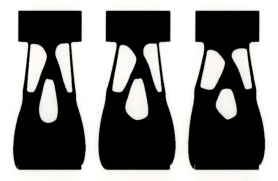

Fig. 5.19: Comparison of results using the set-up from Fig. 5.18. **Left**: Result from Algorithm 4.10 for the expectation-based model. **Middle**: Result from Algorithm 4.10 for the expected excess model (3.42). **Right**: Result obtained for the expected excess model (3.40) with the barrier approach and Algorithm 4.11.

5.2 Risk Aversion

This is seemingly a lot smaller than the one obtained for the other model (3.42) shown in Fig. 5.20, which is 0.12367. However, these values should not be compared directly like that, as they are different objective functionals after all. If we compute the objective value of model (3.42) for the shape obtained by the barrier method given in Fig. 5.21 on the right, we get a value of 0.1353209, which is even slightly bigger than 0.12367.

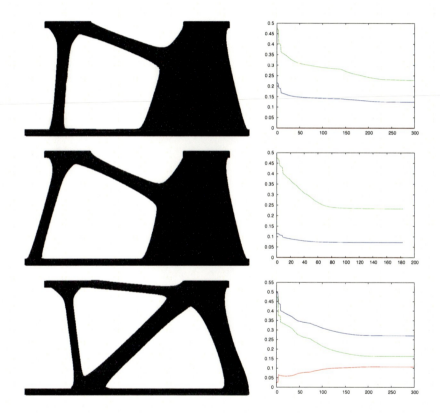

Fig. 5.20: Here we compare the results of Algorithm 4.10 for the approximating expected excess model (3.42) for $\eta = 0.4$ (top) and $\eta = 0.6$ (middle). At the bottom, we show the result obtained from solving the expectation-based problem. The stochastic configuration is like the one in Fig. 5.12. Here, however, the first five scenarios, which correspond to the surface loads acting on the left upper plate, occur with probability 0.01 each, whereas the last five, i.e. the ones acting on the right upper plate, occur with probability 0.19 each.

Fig. 5.21: For comparison, we also ran Algorithm 4.11 using the barrier approach for the same configuration shown in Fig. 5.20 with $\eta = 0.4$. The left picture shows the result with $s = 0.8$ in Algorithm 4.5, whereas the right picture was obtained with $s = 0.9$. The objective value, i.e. the value of the objective function of problem (3.39), for the left shape was 0.00395, and for the right shape 0.00808.

Finally, we consider the set-up from Fig. 5.12 again, where this time the first five scenarios are more likely with probability 0.15 each, whereas the last five occur with probability 0.05 each. As a threshold value we used $\eta = 0.1$. α was set to 0.5, and for Algorithm 4.5 we used $s = 0.8$ and $n_{\text{top}} = 5$. The results are shown in Fig. 5.22 for the expected excess, the excess probability, and the expectation-based objectives. The results clearly differ significantly.

5.2 Risk Aversion

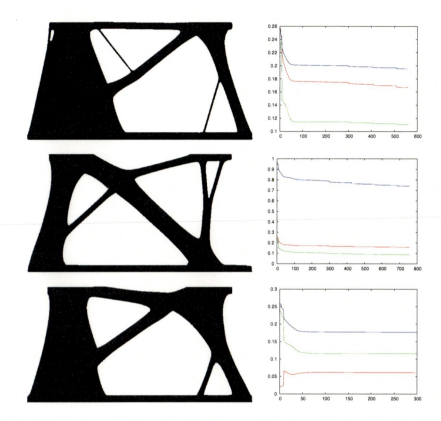

Fig. 5.22: Again, we consider essentially the same stochastic configuration as in Fig. 5.20. This time, scenarios $1,\ldots,5$ occur with probability $\pi_i = 0.15$ each, and the last five with $\pi = 0.05$ each. The top row shows the result from minimizing the expected excess, and the middle row shows the result obtained from minimizing the excess probability. In the last row, we added for comparison the solution to the expectation-based problem. In all energy plots, the red line shows the progression of the expectation of the compliance over the course of iterations.

A Appendix

A.1 Notation

Notation A.1. *We use the following notations for matrices:*

(i) $\mathbb{R}^{m \times n}$ = *space of real $m \times n$ matrices*

(ii) *We write $A = (a_{ij}) \in \mathbb{R}^{m \times n}$ to mean A is an $m \times n$ matrix with $(i, j)^{th}$ entry a_{ij}. Occasionally, we also denote the entry a_{ij} by $(A)_{ij}$.*

(iii) $\text{tr} A$ = *trace of the matrix A.*

(iv) $\det A$ = *determinant of the matrix A.*

(v) A^T = *transpose of the matrix A.*

(vi) *Let $A = (a_{ij})$ and $B = (b_{ij})$ be $m \times n$ matrices. Then the following defines an inner product:*

$$A : B = \text{tr} A^T B = \sum_{i=1}^{m} \sum_{j=1}^{n} a_{ij} b_{ij}. \tag{A.1}$$

Notation A.2.

(i) $\mathbb{N}$ = *set of nonnegative integers.*

(ii) $\mathbb{R}^n$ = *n-dimensional Euclidean space; $\mathbb{R} = \mathbb{R}^1$.*

(iii) *A point $x \in \mathbb{R}^n$ is $x = (x_1, \ldots, x_n)$. x is always regarded as a column vector, and x^T as a row vector. We also write vectors as $x = (x_i)$ similar to the above matrix notation using their entries.*

(iv) *If $\mathcal{O}$ is a subset of $\mathbb{R}^n$, then $\partial \mathcal{O}$ = boundary of $\mathcal{O}$ and $\overline{\mathcal{O}} = \mathcal{O} \cup \partial \mathcal{O}$ = closure of $\mathcal{O}$. $\text{int}\, \mathcal{O}$ denotes the interior of $\mathcal{O}$.*

(v) *If $x = (x_1, \ldots, x_n)$ and $y = (y_1, \ldots, y_n)$ belong to $\mathbb{R}^n$,*

$$x \cdot y = \sum_{i=1}^{n} a_i b_i, \|x\| = \left(\sum_{i=1}^{n} x_i^2 \right)^{\frac{1}{2}}.$$

(vi) $\delta_{ij} = \begin{cases} 1 & \text{if } i = j, \\ 0 & \text{otherwise} \end{cases}$ *denotes Kronecker's delta.*

(vii) *The ith standard coordinate vector in $\mathbb{R}^n$ is denoted by*

$$\mathbf{e}_i = (\delta_{i1}, \ldots, \delta_{in}).$$

(viii) An expression like $x \geq 0$ for a vector $x \in \mathbb{R}^n$ is to be understood componentwise, i.e., $x_i \geq 0$, $\forall i = 1, \ldots, n$.

(ix) For a set $\mathcal{O} \subseteq \mathbb{R}^n$ and a point $x \in \mathbb{R}^n$, we denote the distance from $\mathcal{O}$ to x by $\mathrm{dist}(\mathcal{O}, x)$.

Notation A.3. *For functions, we use the following notations:*

(i) Let $\mathcal{O} \subseteq \mathbb{R}^n$ and $f\colon \mathcal{O} \to \mathbb{R}$ be a function. Then we write
$$f(x) = f(x_1, \ldots, x_n), \quad \forall x \in \mathcal{O}.$$

(ii) If $\mathcal{O} \subseteq \mathbb{R}^n$ and $f\colon \mathcal{O} \to \mathbb{R}^m$, we write
$$f(x) = (f_1(x), \ldots, f_m(x)), \quad \forall x \in \mathcal{O}.$$
The function f_k is the kth component of f, for $k = 1, \ldots, m$.

(iii) Let Γ be a smooth $(n-1)$-dimensional surface in $\mathbb{R}^n$. Then we write
$$\int_\Gamma f \, ds$$
for the integral of f over Γ with respect to the $(n-1)$-dimensional surface measure.

Notation A.4. *Here we collect notations used for derivatives for functions. Let $\mathcal{O} \subseteq \mathbb{R}^n$ be an open subset of $\mathbb{R}^n$. Assume $f\colon \mathcal{O} \to \mathbb{R}$ and $g\colon \mathcal{O} \to \mathbb{R}^m$ with $m > 1$.*

(i) For the ith partial derivative of f we write
$$\frac{\partial f}{\partial x_i}(x) = \lim_{h \to 0} \frac{f(x + h\mathbf{e}_i) - f(x)}{h},$$
provided this limit exists. Sometimes we write f_{x_i} for $\frac{\partial f}{\partial x_i}$.

(ii) Very often, we also write $f_{,i}$ for $\frac{\partial f}{\partial x_i}$. This notation is particularly convenient for derivatives of vector-valued functions which are defined below.

(iii) Similarly, $\frac{\partial^2 f}{\partial x_i \partial x_j} = f_{,ij}$, etc.

(iv) For time-dependent functions $h(t)$, we also use the notation $\dot{h}(t) := \frac{dh}{dt} h(t)$.

(v) Multi-index notation:

(a) $\alpha = (\alpha_1, \ldots, \alpha_n) \in \mathbb{N}^n$ is called a multi-index of order
$$|\alpha| = \alpha_1 + \cdots + \alpha_n.$$

(b) If α is a multi-index, we define
$$D^\alpha f(x) := \frac{\partial^{|\alpha|} f(x)}{\partial x_1^{\alpha_1} \cdots \partial x_n^{\alpha_n}} = \partial_{x_1}^{\alpha_1} \cdots \partial_{x_n}^{\alpha_n} f(x).$$

A.1 Notation

(c) If $m > 1$, we define
$$D^\alpha g(x) = (D^\alpha g_1, \ldots, D^\alpha g_m) \quad \text{for each multi-index } \alpha.$$

(vi) $\nabla f = (f_{,1}, \ldots, f_{,n})^T = $ gradient vector.

(vii) $\nabla^2 f = \begin{pmatrix} f_{,11} & \cdots & f_{,1n} \\ \vdots & \ddots & \vdots \\ f_{,n1} & \cdots & f_{,nn} \end{pmatrix} = $ Hessian matrix.

(viii) $\Delta f = \sum_{i=1}^n f_{,ii} = \text{tr}(\nabla^2 f) = $ Laplacian of f.

(ix) The jth partial derivative of the ith component of g is denoted by $g_{i,j}$ (cf. (ii)).

(x) $\nabla g = \begin{pmatrix} g_{1,1} & \cdots & g_{1,n} \\ \vdots & \ddots & \vdots \\ g_{m,1} & \cdots & g_{m,n} \end{pmatrix} = $ gradient matrix.

(xi) If $m = n$, we have
$$\text{div } g = \text{tr}(\nabla g) = \sum_{i=1}^n g_{i,i} = \text{ divergence of } g.$$

Notation A.5. Let $\mathcal{O} \subseteq \mathbb{R}^n$ be an open domain. For function spaces we use the following notations:

(i) $C(\mathcal{O}) = \{f \colon \mathcal{O} \to \mathbb{R} : f \text{ is continuous}\}$

(ii) $C^k(\mathcal{O}) = \{f \colon \mathcal{O} \to \mathbb{R} : f \text{ is k-times continuously differentiable}\}$

(iii) $C^\infty(\mathcal{O}) = \{f \colon \mathcal{O} \to \mathbb{R} : f \text{ is infinitely differentiable}\} = \bigcap_{k=0}^\infty C^k(\mathcal{O})$.

(iv) $C_0(\mathcal{O}), C_0^k(\mathcal{O})$ are those functions in $C(\mathcal{O}), C^k(\mathcal{O})$ with compact support.

(v) $C^{0,1}(\mathcal{O})$ denotes the set of all Lipschitz continuous functions $f \colon \mathcal{O} \to \mathbb{R}$.

(vi) $L^p(\mathcal{O}) = \{f \colon \mathcal{O} \to \mathbb{R} : f \text{ is Lebesgue measurable}, \|f\|_{L^p(\mathcal{O})} < \infty\}$, where
$$\|f\|_{L^p(\mathcal{O})} = \left(\int_\mathcal{O} |f|^p \, dx\right)^{\frac{1}{p}}, \quad 1 \leq p < \infty.$$

$L^\infty(\mathcal{O}) = \{f \colon \mathcal{O} \to \mathbb{R} : f \text{ is Lebesgue measurable}, \|f\|_{L^\infty(\mathcal{O})} < \infty\}$, where
$$\|f\|_{L^\infty(\mathcal{O})} = \operatorname*{ess\,sup}_{\mathcal{O}} |f|$$
$$= \inf\{\mu \in \mathbb{R} : |\{f > \mu\}| = 0\}.$$

($|.|$ denotes the n-dimensional Lebesgue measure.)

(vii) $W^{k,p}(\mathscr{O})$ for $k = 0, 1, \ldots$, $1 \leq p \leq \infty$ denote Sobolev spaces. They consist of all L^p-functions such that all derivatives up to the order of k exist in the weak sense and belong to $L^p(\mathscr{O})$. In the special case $p = 2$, we usually write

$$H^k(\mathscr{O}) = W^{k,2}(\mathscr{O}), \quad k = 0, 1, \ldots.$$

Note that $H^0 = L^2(\mathscr{O})$. The definition of weak derivatives and further properties can be found in many books about functional analysis and partial differential equations, e.g. in [Alt02, Bra03, Eva02].

(viii) $H_0^1(\mathscr{O})$ denotes the closure of $C_0^\infty(\mathscr{O})$ in $H^1(\mathscr{O})$. $H_0^1(\mathscr{O})$ is interpreted as comprising those functions $u \in H^1(\mathscr{O})$ such that $u = 0$ on $\partial \mathscr{O}$ (in the sense of traces).

(ix) An inner product in $H^k(\mathscr{O})$ is defined by

$$(u,v)_k := \sum_{|\alpha| \leq k} (\partial^\alpha u, \partial^\alpha v)_0,$$

where $(.,.)_0$ denotes the L^2 inner product. The associated norms are

$$\|u\|_k := \sqrt{(u,u)_k} = \sqrt{\sum_{|\alpha| \leq k} \|\partial^\alpha u\|_0^2},$$

as well as the seminorms

$$|u|_k := \sqrt{\sum_{|\alpha|=k} \|\partial^\alpha u\|_0^2}.$$

(x) The spaces $C(\mathscr{O}; \mathbb{R}^m), L^p(\mathscr{O}; \mathbb{R}^m)$, etc. consist of those functions $f \colon \mathscr{O} \to \mathbb{R}^m$, $f = (f_1, \ldots, f_m)$, with $f_i \in C(\mathscr{O}), L^p(\mathscr{O})$, etc. (for all $i = 1, \ldots, m$). For theses spaces we use the same notation for inner products and norms as in the scalar case. For example, if $u, v \in L^2(\mathscr{O}; \mathbb{R}^2), u = (u_1, u_2), v = (v_1, v_2)$ then it is easy to check that $(u,v)_0 := (u_1, v_1)_0 + (u_2, v_2)_0$ defines an inner product on $L^2(\mathscr{O}; \mathbb{R}^2)$, and $\|u\|_0 := \sqrt{(u,u)_0}$ is the associated norm.

A.2 Important Facts and Theorems

Theorem A.6 (Lax-Milgram Theorem, see e.g. [Eva02, p. 297]). *Let H be a real Hilbert space with norm $\|\cdot\|$ and inner product $(\cdot, \cdot)$. The pairing of H and its dual space is denoted by $\langle \cdot, \cdot \rangle$. Assume that*

$$B \colon H \times H \to \mathbb{R}$$

is a bilinear mapping, for which there exist constants $\alpha, \beta > 0$ such that

$$|B(u,v)| \leq \alpha \|u\| \|v\|, \quad \forall u, v \in H,$$

and

$$\beta \|u\|^2 \leq B(u,u), \quad \forall u \in H.$$

A.2 Important Facts and Theorems

Finally, let $f: H \to \mathbb{R}$ be a bounded linear functional on H. Then there exists a unique element $u \in H$ such that

$$B(u,v) = \langle f,v \rangle, \quad \forall v \in H.$$

Theorem A.7 (Korn's second inequality, cf. [Bra03, p. 281]). *Let $\mathcal{O} \subseteq \mathbb{R}^n$ be an open, bounded domain with piecewise smooth boundary[1]. Additionally let $\Gamma_0 \subseteq \partial \mathcal{O}$ have a positive $n-1$-dimensional measure. Then there exists a positive $c' = c'(\mathcal{O}, \Gamma_0)$ such that*

$$\int_{\mathcal{O}} e(v) : e(v) \, dx \geq c' \|v\|_1^2, \quad \forall v \in H^1_{\Gamma_0}(\mathcal{O}; \mathbb{R}^n).$$

$H^1_{\Gamma_0}(\mathcal{O}; \mathbb{R}^n)$ *denotes the closure of* $\{v \in C^\infty(\mathcal{O}; \mathbb{R}^n) : v(x) = 0 \text{ for } x \in \Gamma_0\}$ *with respect to the norm* $\|\cdot\|_1$.

Theorem A.8 (Integration-by-parts formula, cf. [Alt02, p. 252] and [Eva02, p. 268]). *Let $\mathcal{O} \subseteq \mathbb{R}^n$ be open and bounded with Lipschitz boundary. Suppose that $1 \leq p \leq \infty$, $u \in W^{1,p}(\mathcal{O})$, and $v \in W^{1,p'}(\mathcal{O})$ with $\frac{1}{p} + \frac{1}{p'} = 1$. Then the following holds for all $i = 1, \ldots, n$:*

$$\int_{\mathcal{O}} u_{,i} v \, dx = -\int_{\mathcal{O}} u v_{,i} \, dx + \int_{\partial \mathcal{O}} u v \nu_i \, ds, \tag{A.2}$$

where ν denotes the outward pointing unit normal vector field along $\partial \mathcal{O}$.

Theorem A.9 (Coarea formula, cf. [Eva02, p. 629]). *Let $\phi : \mathbb{R}^n \to \mathbb{R}$ be Lipschitz continuous and assume that for almost every $r \in \mathbb{R}$ the level set*

$$\{x \in \mathbb{R}^n : \phi(x) = r\}$$

is a smooth, $(n-1)$-dimensional hypersurface in $\mathbb{R}^n$. Suppose also $f : \mathbb{R}^n \to \mathbb{R}$ is continuous and summable. Then

$$\int_{\mathbb{R}^n} f \|\nabla \phi\| \, dx = \int_{-\infty}^{\infty} \left(\int_{\{\phi=r\}} f \, ds \right) dr.$$

Theorem A.10 (Cauchy-Schwarz inequality, cf. [Alt02, Eva02, Heu06]). *Let X be a pre-Hilbert space with inner product $(\cdot,\cdot)$ and norm $\|\cdot\| = \sqrt{(\cdot,\cdot)}$. Then*

$$|(x,y)| \leq \|x\| \|y\|,$$

for all $x, y \in X$.

Theorem A.11 (Duality Theorem of Linear Programming, see [Bor01, Chv83, Dan74, Sch98]). *Let $A \in \mathbb{R}^{m \times n}$ be a matrix, and $b \in \mathbb{R}^m, c \in \mathbb{R}^n$ be vectors such that $\{x \in \mathbb{R}^n : Ax \leq b\} \neq \emptyset$ and $\{y \in \mathbb{R}^m : A^T y = c, y \geq 0\} \neq \emptyset$. Then it holds that*

$$\underbrace{\max \left\{ c^T x : Ax \leq b \right\}}_{\text{primal problem}} = \underbrace{\min \left\{ b^T y : A^T y = c, y \geq 0 \right\}}_{\text{dual problem}}.$$

---
[1] This theorem also holds for domains with Lipschitz boundary (cf. [Alt02, p. 242]), see e.g. [Tie99].

References

[AA06] AMSTUTZ, SAMUEL and HEIKO ANDRÄ: *A new algorithm for topology optimization using a level-set method.* Journal of Computational Physics, 216:573–588, 2006.

[AB93] AMBROSIO, LUIGI and GIUSEPPE BUTTAZZO: *An optimal design problem with perimeter penalization.* Calc. Var., 1:55–69, 1993.

[AdGJT05] ALLAIRE, GRÉGOIRE, FRÉDÉRIC DE GOURNAY, FRANÇOIS JOUVE and ANCA-MARIA TOADER: *Structural optimization using topological and shape sensitivity via a level set method.* Control and Cybernetics, 34:59–80, 2005.

[AJ05] ALLAIRE, GRÉGOIRE and FRANÇOIS JOUVE: *A level-set method for vibration and multiple loads structural optimization.* Comput. Methods Appl. Mech. Engrg., 194:3269–3290, 2005.

[AJT04] ALLAIRE, GRÉGOIRE, FRANÇOIS JOUVE and ANCA-MARIA TOADER: *Structural Optimization using Sensitivity Analysis and a Level-Set Method.* Journal of Computational Physics, 194(1):363 – 393, 2004.

[All02] ALLAIRE, GRÉGOIRE: *Shape Optimization by the Homogenization Method*, volume 146. Springer Applied Mathematical Sciences, 2002.

[Alt02] ALT, HANS WILHELM: *Lineare Funktionalanalysis: Eine anwendungsorientierte Einführung.* Springer, 2002.

[AP06] ALLAIRE, GRÉGOIRE and O. PANTZ: *Structural optimization with FreeFem++.* Structural and Multidisciplinary Optimization, 32(3):173–181, 2006.

[AS99] ADALSTEINSSON, D. and J. A. SETHIAN: *The Fast Construction of Extension Velocities in Level Set Methods.* Journal of Computational Physics, 148:2–22, 1999.

[BB05] BUCUR, DORIN and GIUSEPPE BUTTAZZO: *Variational Methods in Shape Optimization Problems.* Progress in Nonlinear Differential Equations and Their Applications, Birkhäuser, 2005.

[BDM91] BUTTAZZO, GIUSEPPE and GIANNI DAL MASO: *Shape Optimization for Dirichlet Problems: Relaxed Formulation and Optimality Conditions.* Applied Mathematics and Optimization, 23:17–49, 1991.

[BDM93] BUTTAZZO, GIUSEPPE and GIANNI DAL MASO: *An Existence Result for a Class of Shape Optimization Problems.* Arch. Rational Mech. Anal., 122:183–195, 1993.

[Bea55] BEALE, E. M. L.: *On Minimizing a Convex Function Subject to Linear Inequalities.* J. Royal Statistical Society, Series B, 17:173–184, 1955.

[BGLS03] BONNANS, J. FRÉDÉRIC, J. CHARLES GILBERT, CLAUDE LEMARÉCHAL and CLAUDIA A. SAGASTIZÁBAL: *Numerical Optimization: Theoretical and Practical Aspects.* Springer, 2003.

[BHR04] BURGER, MARTIN, BENJAMIN HACKL and WOLFGANG RING: *Incorporating Topological Derivatives into Level Set Methods.* J. Comp. Phys., 194:344–362, 2004.

[BL97] BIRGE, JOHN R. and FRANÇOIS LOUVEAUX: *Introduction to Stochastic Programming.* Springer Series in Operations Research, 1997.

[BN95] BEHNEN, KONRAD and GEORG NEUHAUS: *Grundkurs Stochastik.* Teubner Stuttgart, 1995.

[BO05] BURGER, MARTIN and STANLEY J. OSHER: *A Survey on Level Set Methods for Inverse Problems and Optimal Design.* European Journal of Applied Mathematics, 16:263–301, 2005.

[Bor01] BORGWARDT, KARL HEINZ: *Optimierung, Operations Research, Spieltheorie: Mathematische Grundlagen.* Birkhäuser, 2001.

[Bra03] BRAESS, DIETRICH: *Finite Elemente: Theorie, schnelle Löser und Anwendungen in der Elastizitätstheorie.* Springer, 2003.

[BS03] BENDSØE, MARTIN P. and OLE SIGMUND: *Topology Optimization: Theory, Methods and Applications.* Springer, 2003.

[BTKNZ99] BEN-TAL, AHARON, M. KOČVARA, A. NEMIROVSKI and J. ZOWE: *Free material sesign via semidefinite programming: the multiload case with contact conditions.* SIAM J. Optim., 9(4):813–832, 1999.

[BTN02] BEN-TAL, AHARON and ARKADI NEMIROVSKI: *Robust Optimization — methodology and applications.* Mathematical Programming, Ser. B, 92(3):453–480, 2002.

[Buc05] BUCUR, DORIN: *How to prove existence in shape optimization.* Control and Cybernetics, 34(1):103–116, 2005.

[Bur03] BURGER, MARTIN: *A framework for the construction of level set methods for shape optimization and reconstruction.* Interfaces and Free Boundaries, 5:301–329, 2003.

[But98] BUTTAZZO, GIUSEPPE: *On the existence of minimizing domains for some shape optimization problems.* In *ESAIM: Proceedings*, volume 3, pages 51–64, 1998.

References

[CC99] CHERKAEV, ANDREJ and ELENA CHERKAEV: *Stable optimal design for uncertain loading conditions*. In AL., V. BERDICHEVSKY ET (editor): *Homogenization*, volume 50 of *Series on Advances in Mathematics for Applied Sciences*, pages 193–213. World Scientific, Singapore, 1999.

[CC03] CHERKAEV, ANDREJ and ELENA CHERKAEV: *Principal Compliance and Robust Optimal Design*. Journal of Elasticity, 72:71–98, 2003.

[Cha03] CHAMBOLLE, ANTONIN: *A Density Result in Two-Dimensional Elasticity, and Applications*. Arch. Rational Mech. Anal., 167:211–233, 2003.

[CHP+09] CONTI, SERGIO, HARALD HELD, MARTIN PACH, MARTIN RUMPF and RÜDIGER SCHULTZ: *Shape Optimization under Uncertainty - A Stochastic Programming Perspective*. SIAM J. Optim., 19(4):1610–1632, 2009.

[Chv83] CHVÁTAL, VAŠEK: *Linear Programming*. Freeman, 1983.

[Cia88] CIARLET, PHILIPPE G.: *Mathematical Elasticity Volume I: Three-Dimensional Elasticity*, volume 20. Studies in Mathematics and its Applications, North-Holland, 1988.

[Dan55] DANTZIG, GEORGE B.: *Linear programming under uncertainty*. Management Science, 1:197–206, 1955.

[Dan74] DANTZIG, GEORGE BERNARD: *Linear Programming and Extensions*. Princeton University Press, 6th edition, 1974.

[Das] DASHOPTIMIZATION: *Xpress-MP*. http://www.dashoptimization.com/.

[dG06] GOURNAY, FRÉDÉRIC DE: *Velocity Extension for the Level-Set Method and Multiple Eigenvalues in Shape Optimization*. SIAM Journal on Control and Optimization, 45(1):343–367, 2006.

[dGAJ06] GOURNAY, FRÉDÉRIC DE, GRÉGOIRE ALLAIRE and FRANÇOIS JOUVE: *Shape and topology optimization of the robust compliance via the level set method*. To appear in ESAIM: Control, Optimisation and Calculus of Variations, 2006.

[DJPZ01] DELFOUR, M. C. and J.-P. ZOLÉSIO: *Shapes and Geometries: Analysis, Differential Calculus, and Optimization*. Siam, 2001.

[DR75] DAVIS, PHILIP J. and PHILIP RABINOWITZ: *Methods of Numerical Integration*. Computer Science and Applied Mathematics. Academic Press, 1975.

[DS57] DUNFORD, NELSON and JACOB T. SCHWARTZ: *Linear Operators Part I: General Theory*. Interscience Publishers, INC., New York, 1957.

[Ehr05] EHRGOTT, MATTHIAS: *Multicriteria Optimization*. Springer, 2nd edition, 2005.

[Els02] ELSTRODT, JÜRGEN: *Maß- und Integrationstheorie*. Grundwissen Mathematik. Springer, 3. edition, 2002.

[ET76]   EKELAND, IVAR and ROGER TEMAM: *Convex Analysis and Variational Problems*, volume 1. Studies in Mathematics and its Applications, North-Holland, 1976.

[Eva02]   EVANS, LAWRENCE C.: *Partial Differential Equations*, volume 19. AMS Graduate Studies in Mathematics, 2002.

[FLSS07] FULMAŃSKI, PIOTR, ANTOINE LAURAIN, JEAN-FRANCOIS SCHEID and JAN SOKOŁOWSKI: *A Level Set Method in Shape and Topology Optimization For Variational Inequalities*. Int. J. Appl. Math. Comput. Sci., 17(3):413–430, 2007.

[GGM01]   GARREAU, STÉPHANE, PHILIPPE GUILLAUME and MOHAMED MASMOUDI: *The Topological Asymptotic for PDE Systems: The Elasticity Case*. SIAM J. Control Optim., 39(6):1756–1778, 2001.

[GK02]   GEIGER, CARL and CHRISTIAN KANZOW: *Theorie und Numerik restringierter Optimierungsaufgaben*. Springer, 2002.

[Hac85]   HACKBUSCH, WOLFGANG: *Multi-Grid Methods and Applications*. Springer Series in Computational Mathematics, 1985.

[Heu06]   HEUSER, HARRO: *Funktionalanalysis*. Teubner, 2006.

[HHW05]   HELD, HARALD, RAYMOND HEMMECKE and DAVID L. WOODRUFF: *A Decomposition Algorithm Applied to Planning the Interdiction of Stochastic Networks*. Naval Research Logistics, 52(4):321–328, 2005.

[HKO07]   HE, LIN, CHIU-YEN KAO and STANLEY OSHER: *Incorporating topological derivatives into shape derivatives based level set methods*. Journal of Computational Physics, 225(1):891–909, 2007.

[HL07]   HINTERMÜLLER, M. and A. LAURAIN: *Where to place a hole?* European Consortium for Mathematics in Industry, ECMI Newsletter 41, 2007.

[HN97]   HASLINGER, J. and P. NEITTAANMÄKI: *Finite Element Approximation for Optimal Shape, Material and Topology Design*. Wiley, 1997.

[HS97a]   HACKBUSCH, WOLFGANG and STEFAN A. SAUTER: *Composite finite elements for problems containing small geometric details - Part II: Implementation and numerical results*. Computing and Visualization in Science, 1(1):15–25, 1997.

[HS97b]   HACKBUSCH, WOLFGANG and STEFAN A. SAUTER: *Composite finite elements for the approximation of PDEs on domains with complicated microstructures*. Numer. Math., 75:447–472, 1997.

[Huy01]   HUYSE, LUC.: *Free-form airfoil shape optimization under uncertainty using maximum expected value and second-order second-moment strategies*. ICASE report ; no. 2001-18. ICASE, NASA Langley Research Center Available from NASA Center for Aerospace Information, Hampton, VA, 2001.

References

[ILO]   ILOG: *ILOG CPLEX*. http://www.ilog.com/products/cplex/.

[KW94]   KALL, PETER and STEIN W. WALLACE: *Stochastic Programming*. Wiley-Interscience Series in Systems and Optimization, 1994.

[LPR+07]   LIEHR, FLORIAN, TOBIAS PREUSSER, MARTIN RUMPF, STEFAN SAUTER and LARS OLE SCHWEN: *Composite Finite Elements for 3D Image Based Computing*. Computing and Visualization in Science, 2007.

[LS03]   LOUVEAUX, FRANÇOIS and RÜDIGER SCHULTZ: *Stochastic Integer Programming*. In RUSZCZYŃSKI, ANDRZEJ and ALEXANDER SHAPIRO (editors): *Stochastic Programming*, volume 10 of *Handbooks in Operations Research and Management Science*, pages 213–266. Elsevier Science, 2003.

[Mak]   MAKHORIN, ANDREW O.: *GLPK (GNU Linear Programming Kit)*. http://www.gnu.org/software/glpk/.

[MS00]   MAAR, B. and V. SCHULZ: *Interior point multigrid methods for topology optimization*. Struct Multidisc Optim, 19:214–224, 2000.

[NP02]   NOVRUZI, ARIAN and MICHEL PIERRE: *Structure of shape derivatives*. Journal of Evolution Equations, 2:365–382, 2002.

[NR]   NOVRUZI, ARIAN and JEAN R. ROCHE: *Second Order Derivatives, Newton Method, Application to Shape Optimization*. http://citeseer.ist.psu.edu/novruzi95second.html.

[NW99]   NOCEDAL, JORGE and STEPHEN J. WRIGHT: *Numerical Optimization*. Springer Series in Operations Research, 1999.

[OF03]   OSHER, STANLEY and RONALD FEDKIW: *Level Set Methods and Dynamic Implicit Surfaces*, volume 153. Applied Mathematical Sciences, Springer, 2003.

[OS88]   OSHER, STANLEY and JAMES A. SETHIAN: *Fronts Propagating with Curvature-Dependent Speed: Algorithms Based on Hamilton–Jacobi Formulations*. Journal of Computational Physics, 79(1):12–49, 1988.

[Pac05]   PACH, MARTIN: *Levelsetverfahren in der Shapeoptimierung*. University of Duisburg-Essen, 2005. Diploma thesis, available here: http://www.uni-duisburg.de/FB11/disma/m_pach/ShapeOpt.pdf.

[PM02]   PHILPOTT, ANDY and ANDREW MASON: *Advances In Optimization In Yacht Performance Analysis*. http://citeseer.ist.psu.edu/philpott02advances.html, 2002.

[Pré95]   PRÉKOPA, ANDRÁS: *Stochastic Programming*. Kluwer, 1995.

[RS01]   RÖMISCH, WERNER and RÜDIGER SCHULTZ: *Multistage stochastic integer programming: an introduction*. In GRÖTSCHEL, M., S. O. KRUMKE and J. RAMBAU (editors): *Online Optimization of Large Scale Systems*, pages 581 – 600. Springer, 2001.

References

[RS03a] RIIS, MORTEN and RÜDIGER SCHULTZ: *Applying the Minimum Risk Criterion in Stochastic Recourse Programs.* Computational Optimization and Applications, 24:267–287, 2003.

[RS03b] RUSZCZYŃSKI, ANDRZEJ and ALEXANDER SHAPIRO: *Stochastic Programming Models.* In RUSZCZYŃSKI, ANDRZEJ and ALEXANDER SHAPIRO (editors): *Stochastic Programming*, volume 10 of *Handbooks in Operations Research and Management Science*, pages 1–64. Elsevier Science, 2003.

[RSS06] RECH, M., S. SAUTER and A. SMOLIANSKI: *Two-Scale Composite Finite Element Method for the Dirichlet Problem on Complicated Domains.* Numer. Math., 102(4):681–708, 2006.

[Rus99] RUSZCZYŃSKI, ANDRZEJ: *Some advances in decomposition methods for stochastic linear programming.* Annals of Operations Research, 85:153–172, 1999.

[Rus06] RUSZCZYŃSKI, ANDRZEJ: *Nonlinear Optimization.* Princeton University Press, 2006.

[Sau02] SAUTER, STEFAN: *Composite Finite Elements and Multigrid Lecture Notes of the Zürich Summerschool 02.* Preprint 22, Universität Zürich, 2002.

[Sch96] SCHUMACHER, AXEL: *Topologieoptimierung von Bauteilstrukturen unter Verwendung von Lochpositionierungskriterien.* PhD thesis, Universität – Gesamthochschule Siegen, 1996.

[Sch98] SCHRIJVER, ALEXANDER: *Theory of Linear and Integer Programming.* Wiley-Interscience series in discrete mathematics and optimization. Wiley, 1998.

[Sch03a] SCHULTZ, RÜDIGER: *Mixed-integer value functions in stochastic programming.* In *Combinatorial Optimization - Eureka, You Shrink! Papers Dedicated to Jack Edmonds*, number 2570 in *Lecture Notes in Computer Science*, pages 171–184. Springer, 2003.

[Sch03b] SCHULTZ, RÜDIGER: *Stochastic programming with integer variables.* Mathematical Programming, 97:285–309, 2003.

[Sch05] SCHULTZ, RÜDIGER: *Risk Aversion in Two-Stage Stochastic Integer Programming.* Preprint 612-2005, Department of Mathematics, University of Duisburg-Essen, 2005.

[Set01] SETHIAN, J. A.: *Evolution, implementation, and application of level set and fast marching methods for advancing fronts.* J. Comput. Phys., 169(2):503–555, 2001.

[Soi03] SOILLE, PIERRE: *Morphological Image Analysis: Principles and Applications.* Springer, 2nd edition, 2003.

[SS03] STOLPE, M. and K. SVANBERG: *Modelling topology optimization problems as linear mixed 0-1 programs.* Int. J. Numer. Meth. Engng, 57:723–739, 2003.

References

[ST06] SCHULTZ, RÜDIGER and STEPHAN TIEDEMANN: *Conditional Value-at-Risk in Stochastic Programs with Mixed-Integer Recourse.* Springer Mathematical Programming Series B, 105:365–386, 2006.

[Sta08] STANGL, CLAUDIA: *Strukturoptimierung mit IPOPT im deterministischen und stochastischen Fall.* University of Duisburg-Essen, 2008. Diploma thesis.

[Str71] STROUD, A. H.: *Approximate Calculation of Multiple Integrals.* Series in Automatic Computation. Prentice-Hall, 1971.

[SZ92] SOKOŁOWSKI, JAN and JEAN-PAUL ZOLÉSIO: *Introduction to Shape Optimization: Shape Sensitivity Analysis.* Springer, 1992.

[SZ99] SOKOŁOWSKI, JAN and A. ŻOCHOWSKI: *On the Topological Derivative in Shape Optimization.* SIAM J. Control Optim., 37(4):1251–1272, 1999.

[SZ01] SOKOŁOWSKI, JAN and ANTONI ŻOCHOWSKI: *Topological Derivatives of Shape Functionals for Elasticity Systems.* Mech. Struct. & Mach., 29(3):331–349, 2001.

[SZ03] SOKOŁOWSKI, JAN and ANTONI ŻOCHOWSKI: *Optimality Conditions for Simultaneous Topology and Shape Optimization.* SIAM Journal on Control and Optimization, 42(4):1198–1221, 2003.

[Tie99] TIERO, ALESSANDRO: *On Korn's Inequality in the Second Case.* Journal of Elasticity, 54(3):187–191, 1999.

[Tie05] TIEDEMANN, STEPHAN: *Risk Measures with Preselected Tolerance Levels in Tow-Stage Stochastic Mixed-Integer Programming.* Cuvillier Verlag Göttingen, 2005.

[Trö05] TRÖLTZSCH, FREDI: *Optimale Steuerung partieller Differentialgleichungen.* Vieweg, 2005.

[Š93] ŠVERÁK, V.: *On Optimal Shape Design.* J. Math. Pures Appl., 72:537–551, 1993.

[VSW69] VAN SLYKE, RICHARD and ROGER WETS: *L-shaped linear programs with application to optimal control and stochastic programming.* SIAM Journal on Applied Mathematics, 17(4), 1969.

[War03] WARNKE, RAINER: *Schnelle Löser für elliptische Randwertprobleme mit springenden Koeffizienten.* PhD thesis, Universität Zürich, 2003.

[WB06] WÄCHTER, A. and L. T. BIEGLER: *On the Implementation of a Primal-Dual Interior Point Filter Line Search Algorithm for Large-Scale Nonlinear Programming.* Mathematical Programming, 106(1):25–57, 2006.

[Wol98] WOLSEY, LAURENCE A.: *Integer Programming.* Wiley-Interscience series in discrete mathematics and optimization. Wiley, 1998.

[WZ05] WALLACE, STEIN W. and WILLIAM T. ZIEMBA: *Applications of Stochastic Programming*, volume MPS-SIAM Series on Optimization. SIAM and MPS, 2005.

[Ye97]   YE, YINYU: *Interior Point Algorithms: Theory and Analysis*. Wiley-Interscience Series in Discrete Mathematics and Optimization, 1997.

[ZCMO96]   ZHAO, HONG-KAI, T. CHAN, B. MERRIMAN and S. OSHER: *A Variational Level Set Approach to Multiphase Motion*. Journal of Computational Physics, 127:179–195, 1996.

[Zie98]   ZIEGLER, GÜNTER M.: *Lectures on polytopes*, volume 152 of *Graduate texts in mathematics*. Springer, corr. 2. print. edition, 1998.